T0344679

BASIC STRUCTURAL DYNAMICS

BASIC STRUCTURAL DYNAMICS

James C. Anderson Ph.D.
Professor of Civil Engineering,
University of Southern California

Farzad Naeim Ph.D., S.E., Esq.
Vice President and General Counsel,
John A. Martin & Associates, Inc.

WILEY
JOHN WILEY & SONS, INC.

Published by John Wiley & Sons, Inc., Hoboken, New Jersey.
Published simultaneously in Canada.

For general information about our other products and services, please contact our Customer Care
Department within the United States at (800) 762-2974, outside the United States at (317)
572-3993 or fax (317) 572-4002.

Wiley publishes in a variety of print and electronic formats and by print-on-demand. Some
material included with standard print versions of this book may not be included in e-books or in
print-on-demand. If this book refers to media such as a CD or DVD that is not included in the
version you purchased, you may download this material at http://booksupport.wiley.com. For more
information about Wiley products, visit www.wiley.com.

Library of Congress Cataloging-in-Publication Data:

Anderson, J. C. (James C.), 1939-
  Basic structural dynamics / James C. Anderson, Farzad Naeim.
    pages cm
  Includes bibliographical references and index.
  ISBN: 978-0-470-87939-9; 978-111-827908-3 (ebk); 978-111-827909-0 (ebk); 978-111-827910-6
(ebk); 978-111-827911-3 (ebk); 978-111-827912-0 (ebk); 978-111-827913-7 (ebk)
  1. Structural dynamics–Textbooks. I. Naeim, Farzad. II. Title.
  TA654.A65 2012
  624.1'71–dc23

                                                                        2012013717

Printed in the United States of America

10 9 8 7 6 5 4 3 2 1

To our wives, Katherine and Fariba

CONTENTS

PREFACE

Our experience of over 30 years of teaching structural dynamics has demonstrated to us that, more often than not, novice students of structural dynamics find the subject foreign and difficult to understand. The main objective of this book is to overcome this hurdle and provide a textbook that is easy to understand and relatively short—a book that can be used as an efficient tool for teaching a first course on the subject without overwhelming the students who are just beginning their study of structural dynamics. There is no shortage of good and comprehensive textbooks on structural dynamics, and once the student has mastered the basics of the subject, he or she can more efficiently navigate the more complex and intricate subjects in this field. This book may also prove useful as a reference for practicing engineers who are not familiar with structural dynamics or those who want a better understanding of the various code provisions that are based on the dynamic response of structures and/or components.

This book is also perhaps unique in that it integrates MATLAB applications throughout. Example problems are generally worked by hand and then followed by MATLAB algorithms and solutions of the same. This will help students solve more problems without getting bogged down in extensive hand calculations that would otherwise be necessary. It will also let students experiment with changing various parameters of a dynamic problem and get a feel for how changing various parameters will affect the outcome. Extensive use is made of the graphics in MAT-LAB to make the concept of dynamic response real. We decided to use MATLAB in many of the examples in the book because (1) it is a very powerful tool, (2) it is easy to use, and (3) a free or nominally priced "student version" is available to virtually all engineering students. We have consciously decided not to include a tutorial on basic MATLAB operations simply because such information is readily available within the help files supplied with MATLAB and in the documentation that is shipped with the student version of MATLAB.

More than 20 years ago, it was decided that, because of the seismic risk in California and the fact that at that time most of our undergraduate students came from California, a course titled "Introduction to Structural Dynamics" was needed. This course was intended for seniors and first-year graduate students in structural engineering. During this time period, much has changed in this important area of study. There has been a tremendous change in both computational hardware and software, which are now readily available to students. Much has also been learned from the occurrence of major earthquakes in various locations around the world and the recorded data that have been obtained from these earthquakes, including both building data and free field data. This book attempts to draw on and reflect these changes to the extent practical and useful to its intended audience.

The book is conceptually composed of three parts. The first part, consisting of Chapters 1 to 6, covers the basic concepts and dynamic response of single-degree-of-freedom systems to various excitations. The second part, consisting of Chapters 7 and 8, covers the linear and non-linear response of multiple-degree-of-freedom systems to various excita-tions. Finally, the third part, consisting of Chapter 9 and the Appendix, deals with the linear and nonlinear response of structures subjected to earthquake ground motions and structural dynamics–related code provi-sions for assessing the seismic response of structures. It is anticipated that for a semester-long introductory course on structural dynamics, Chapters 1 to 7 with selected sections of the other chapters will be covered in the classroom.

This book assumes the student is familiar with at least a first course in differential equations and elementary matrix algebra. Experience with computer programming is helpful but not essential.

JAMES C. ANDERSON, LOS ANGELES, CA
FARZAD NAEIM, LOS ANGELES, CA

CHAPTER 1

BASIC CONCEPTS OF STRUCTURAL DYNAMICS

1.1 THE DYNAMIC ENVIRONMENT

Structural engineers are familiar with the analysis of structures for static loads in which a load is applied to the structure and a single solution is obtained for the resulting displacements and member forces. When considering the analysis of structures for dynamic loads, the term *dynamic* simply means "time-varying." Hence, the loading and all aspects of the response vary with time. This results in possible solutions at each instant during the time interval under consideration. From an engineering standpoint, the maximum values of the structural response are usually the ones of particular interest, especially when considering the case of structural design.

Two different approaches, which are characterized as either deterministic or nondeterministic, can be used for evaluating the structural response to dynamic loads. If the time variation of the loading is fully known, the analysis of the structural response is referred to as a *deterministic analysis*. This is the case even if the loading is highly oscillatory or irregular in character. The analysis leads to a time history of the displacements in the structure corresponding to the prescribed time history of the loading. Other response parameters such as member forces and

1

relative member displacements are then determined from the displacement history.

If the time variation of the dynamic load is not completely known but can be defined in a statistical sense, the loading is referred to as a *random dynamic loading*, and the analysis is referred to as *nondeterministic*. The nondeterministic analysis provides information about the displacements in a statistical sense, which results from the statistically defined loading. Hence, the time variation of the displacements is not determined, and other response parameters must be evaluated directly from an independent nondeterministic analysis rather than from the displacement results. Methods for nondeterministic analysis are described in books on random vibration. In this text, we only discuss methods for deterministic analysis.

1.2 TYPES OF DYNAMIC LOADING

Most structural systems will be subjected to some form of dynamic loading during their lifetime. The sources of these loads are many and varied. The ones that have the most effect on structures can be classified as environmental loads that arise from winds, waves, and earthquakes. A second group of dynamic loads occurs as a result of equipment motions that arise in reciprocating and rotating machines, turbines, and conveyor systems. A third group is caused by the passage of vehicles and trucks over a bridge. Blast-induced loads can arise as the result of chemical explosions or breaks in pressure vessels or pressurized transmission lines.

For the dynamic analysis of structures, deterministic loads can be divided into two categories: periodic and nonperiodic. Periodic loads have the same time variation for a large number of successive cycles. The basic periodic loading is termed *simple harmonic* and has a sinusoidal variation. Other forms of periodic loading are often more complex and nonharmonic. However, these can be represented by summing a sufficient number of harmonic components in a Fourier series analysis. Nonperiodic loading varies from very short duration loads (air blasts) to long-duration loads (winds or waves). An air blast caused by some form of chemical explosion generally results in a high-pressure force having a very short duration (milliseconds). Special simplified forms of analysis may be used under certain conditions for this loading, particularly for design. Earthquake loads that develop in structures as a result of ground motions at the base can have a duration that varies from a few seconds to a few minutes. In this case, general dynamic analysis procedures must be applied. Wind loads are a function of the wind velocity and the height, shape, and stiffness of the structure. These characteristics give rise to aerodynamic forces that can be either calculated or obtained from wind tunnel tests. They are usually represented as equivalent static pressures acting on the surface of the structure.

1.3 BASIC PRINCIPLES

The fundamental physical laws that form the basis of structural dynamics were postulated by Sir Isaac Newton in the *Principia* (1687).[1] These laws are also known as *Newton's laws of motion* and can be summarized as follows:

First law: A particle of constant mass remains at rest or moves with a constant velocity in a straight line unless acted upon by a force.

Second law: A particle acted upon by a force moves such that the time rate of change of its linear momentum equals the force.

Third law: If two particles act on each other, the force exerted by the first on the second is equal in magnitude and opposite in direction to the force exerted by the second on the first.

Newton referred to the product of the mass, m, and the velocity, dv/dt, as the quantity of motion that we now identify as the *momentum*. Then Newton's second law of *linear momentum* becomes

$$\frac{d(m\dot{v})}{dt} = f \tag{1.1}$$

where both the momentum, $m(dv/dt)$, and the driving force, f, are functions of time. In most problems of structural dynamics, the mass remains constant, and Equation (1.1) becomes

$$m\left(\frac{d\dot{v}}{dt}\right) = ma = f \tag{1.2}$$

An exception occurs in rocket propulsion in which the vehicle is losing mass as it ascends. In the remainder of this text, time derivatives will be denoted by dots over a variable. In this notation, Equation (1.2) becomes $m\ddot{v} = f$.

[1] I. Newton, *The Principia: Mathematical Principles of Natural Philosophy*, 1687.

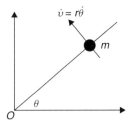

Figure 1.1   Rotation of a mass about a fixed point (F. Naeim, *The Seismic Design Handbook*, 2nd ed. (Dordrecht, Netherlands: Springer, 2001), reproduced with kind permission from Springer Science+Business Media B.V.)

Newton's second law can also be applied to rotational motion, as shown in Figure 1.1. The *angular momentum*, or moment of momentum, about an origin O can be expressed as

$$L = r(m\dot{v}) \tag{1.3}$$

where L = the angular momentum
   r = the distance from the origin to the mass, m
   $\dot{v}$ = the velocity of the mass

When the mass is moving in a circular arc about the origin, the angular speed is $\dot{\theta}$, and the velocity of the mass is $r\dot{\theta}$. Hence, the angular momentum becomes

$$L = mr^2\dot{\theta} \tag{1.4}$$

The time rate of change of the angular momentum equals the torque:

$$\text{torque} = N = \frac{dL}{dt} = mr^2\ddot{\theta} \tag{1.5}$$

If the quantity mr^2 is defined as the moment of inertia, I_θ, of the mass about the axis of rotation (mass moment of inertia), the torque can be expressed as

$$I_\theta \ddot{\theta} = N \tag{1.6}$$

where $d^2\theta/dt^2$ denotes the angular acceleration of the moving mass; in general, $I_\theta = \int \rho^2 dm$. For a uniform material of mass density μ, the mass moment of inertia can be expressed as

$$I_\theta = \mu \int \rho^2 dV \tag{1.7}$$

The rotational inertia about any reference axis, G, can be obtained from the parallel axis theorem as

$$I_G = I_\theta + mr^2 \tag{1.8}$$

Example 1.1   Consider the circular disk shown in Figure 1.2a. Determine the mass moment of inertia of the disk about its center if it has mass density (mass/unit volume) μ, radius r, and thickness t. Also determine the mass moment of inertia of a rectangular rod rotating about one end, as shown in Figure 1.2b. The mass density of the rod is μ, the dimensions of the cross section are $b \times d$, and the length is r.

$$I_0 = \mu \int \rho^2 dV \quad \text{where} \quad dV = \rho(d\theta)(d\rho)t$$

$$I_0 = \mu t \int_0^{2\pi} \int_0^r \rho^3 d\rho d\theta = \mu t \pi \frac{r^4}{2}$$

The mass of the circular disk is $m = \pi r^2 t \mu$.

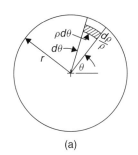

(a)

Figure 1.2a   Circular disk

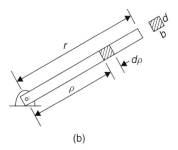

(b)

Figure 1.2b   Rectangular rod

Hence, the mass moment of inertia of the disk becomes

$$I_0 = \frac{mr^2}{2}$$

$$I_0 = \mu \int \rho^2 dV \quad \text{where, for a rectangular rod:} \quad dV = (bd)d\rho$$

$$I_0 = \mu bd \int_0^r \rho^2 d\rho = \mu bd \frac{r^3}{3}$$

The mass of the rod is $m = bdr\mu$, and the mass moment of inertia of the rod becomes

$$I_0 = \frac{mr^2}{3}$$

The rigid-body mass properties of some common structural geometric shapes are summarized in Figure 1.3.

The difference between *mass* and *weight* is sometimes confusing, particularly to those taking a first course in structural dynamics. The mass, m, is a measure of the quantity of matter, whereas the weight, w, is a measure of the force necessary to impart a specified acceleration to a given mass. The acceleration of gravity, g, is the acceleration that the gravity of the earth would impart to a free-falling body at sea level, which is 32.17 ft/sec^2 or 386.1 in/sec^2. For engineering calculations, the acceleration of gravity is often rounded to 32.2 ft/sec^2, which results in 386.4 in/sec^2 when multiplied by 12 in/ft. Therefore, mass does not equal weight but is related by the expression $w = mg$. To keep this concept straight, it is helpful to carry units along with the mathematical operations.

The concepts of the *work* done by a force, and of the *potential and kinetic energies*, are important in many problems of dynamics. Multiply both sides of Equation (1.2) by dv/dt and integrate with respect to time:

$$\int_{t_1}^{t_2} f(t)\dot{v}dt = \int_{t_1}^{t_2} m\ddot{v}\dot{v}dt \tag{1.9}$$

Because $\dot{v}dt = dv$ and $\ddot{v}dt = d\dot{v}$, Equation (1.9) can be written as

$$\int_{v_1}^{v_2} f(t)dv = \frac{1}{2}m\left(\dot{v}_2^2 - \dot{v}_1^2\right) \tag{1.10}$$

The integral on the left side of Equation (1.10) is the area under the force-displacement curve and represents the work done by the force $f(t)$. The two

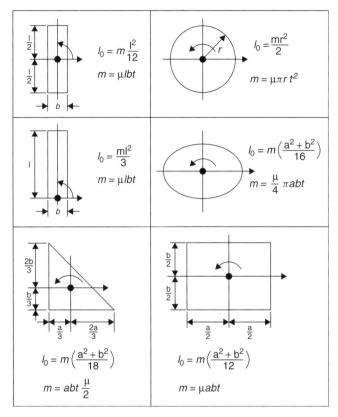

Figure 1.3   Transitional mass and mass moment of inertia (F. Naeim, *The Seismic Design Handbook*, 2nd ed. (Dordrecht, Netherlands: Springer, 2001), reproduced with kind permission from Springer Science+Business Media B.V.)

terms on the right represent the final and initial kinetic energies of the mass. Hence, the work done is equal to the change in kinetic energy.

Consider a force that is acting during the time interval (t_1, t_2). The integral $I = \int f(t)dt$ is defined as the impulse of the force during the time interval. According to Newton's second law of motion, $f = m\dot{v}$. If both sides are integrated with respect to t,

$$I = \int_{t_1}^{t_2} f(t)dt = m(\dot{v}_2 - \dot{v}_1) \qquad (1.11)$$

Hence, the impulse, I, is equal to the change in the momentum. This relation will be useful in analyzing the result of applying a large force for a brief interval of time as will be demonstrated in a later section.

Newton's laws of motion lead, in special circumstances, to the following three important properties of motion (conservation laws):

1. If the sum of the forces acting on a mass is zero, the linear momentum is constant in time.
2. If the sum of the external torques acting on a particle is zero, the angular momentum is constant in time.
3. In a conservative force field, the sum of the kinetic and potential energies remains constant during the motion.

It should be noted that nonconservative forces include frictional forces and forces that depend on velocity and time.

The term *degrees of freedom* in a dynamic system refers to the least number of displacement coordinates needed to define the motion of the system. If the physical system is represented as a continuum, an infinite number of coordinates would be needed to define the position of all the mass of the system. The system would thus have infinitely many degrees of freedom. In most structural systems, however, simplifying assumptions can be applied to reduce the degrees of freedom and still obtain an accurate determination of the displacement.

A *constraint* is a restriction on the possible deformed shape of a system, and a *virtual displacement* is an infinitesimal, imaginary change in configuration that is consistent with the constraints.

In 1717, Johann Bernoulli posed his *principle of virtual work*, which is basically a definition of equilibrium that applies to dynamic as well as static systems. The principle of virtual work states that if, for any arbitrary virtual displacement that is compatible with the system constraints, the virtual work under a set of forces is zero, then the system is in equilibrium. This principle can be restated in terms of virtual displacements—a form that is more applicable to structural systems. It states that if a system that is in equilibrium under a set of forces is subjected to a virtual displacement that is compatible with the system constraints, then the total work done by the forces is zero. The vanishing of the virtual work done is equivalent to a statement of equilibrium.

In his book *Traite de Dynamique* (1743)[2], the French mathematician Jean le Rond d'Alembert proposed a principle that would reduce a problem in dynamics to an equivalent one in statics. He developed the idea

---

[2]J. d'Alembert, *Traite de Dynamique*, 1743, available at http://www.archive.org/details/trait dedynamiqu00dalgoog.

that mass develops an inertia force that is proportional to its acceleration and opposing it:

$$f_i = -m\ddot{v} \tag{1.12}$$

d'Alembert's principle also states that the applied forces together with the forces of inertia form a system in equilibrium.

1.4 DYNAMIC EQUILIBRIUM

The basic equation of static equilibrium used in the displacement method of analysis has the form

$$p = kv \tag{1.13}$$

where p = the applied force
k = the stiffness resistance
v = the resulting displacement

If the statically applied force is now replaced by a dynamic or time-varying force, $p(t)$, the equation of static equilibrium becomes one of dynamic equilibrium and has the form

$$p(t) = m\ddot{v}(t) + c\dot{v}(t) + kv(t) \tag{1.14}$$

where the dot represents differentiation with respect to time.

A direct comparison of these two equations indicates that two significant changes that distinguish the static problem from the dynamic problem were made to Equation (1.13) in order to obtain Equation (1.14). First, the applied load and the resulting response are now functions of time; hence, Equation (1.14) must be satisfied at each instant of time during the time interval under consideration. For this reason, it is usually referred to as an *equation of motion*. Second, the time dependence of the displacements gives rise to two additional forces that resist the applied force and have been added to the right side.

The first term is based on Newton's second law of motion and incorporates d'Alembert's concept of an inertia force that opposes the motion. The second term accounts for dissipative or damping forces that are inferred from the observed fact that oscillations in a structure tend to diminish with time once the time-dependent applied force is removed. These forces are generally represented by viscous damping forces that are proportional to the velocity with the constant of proportionality, c, referred to as the *damping coefficient*:

$$f_d = c\dot{v} \tag{1.15}$$

It must also be recognized that all structures are subjected to gravity loads such as self-weight (dead load) and occupancy load (live load) in addition to any dynamic loading. In an elastic system, the principle of superposition can be applied, so that the responses to static and dynamic loadings can be considered separately and then combined to obtain the total structural response. However, if the structural behavior becomes nonlinear, the response becomes dependent on the load path, and the gravity loads must be considered concurrently with the dynamic loading.

Under the action of severe dynamic loading, the structure will most likely experience nonlinear behavior, which can be caused by material nonlinearity and/or geometric nonlinearity. Material nonlinearity occurs when stresses at certain critical regions in the structure exceed the elastic limit of the material. The equation of dynamic equilibrium for this case has the general form

$$p(t) = m\ddot{v}(t) + c\dot{v}(t) + k(t)v(t) \tag{1.16}$$

where the stiffness or resistance, k, is a function of the yield condition in the structure, which, in turn, is a function of time. Geometric nonlinearity is caused by the gravity loads acting on the deformed position of the structure. If the lateral displacements are small, this effect, which is often referred to as the $P\text{-}\Delta$ *effect*, can be neglected. However, if the lateral displacements become large, this effect must be considered by augmenting the stiffness matrix, k, with the geometric stiffness matrix, k_g, which includes the effect of axial loads.

In order to define the inertia forces completely, it would be necessary to consider the acceleration of every mass particle in the structure and the corresponding displacement. Such a solution would be prohibitively complicated and time-consuming. The analysis procedure can be greatly simplified if the mass of the structure can be concentrated (lumped) at a finite number of discrete points and the dynamic response of the structure can be represented in terms of this limited number of displacement components (degrees of freedom). The number of degrees of freedom required to obtain an adequate solution will depend on the complexity of the structural system. For some structures, a single degree of freedom may be sufficient, whereas, for others, several hundred degrees of freedom may be required.

PROBLEMS

Problem 1.1

Determine the mass moment of inertia of the rectangular and triangular plates when they rotate about the hinges, as shown in Figure 1.4. Assume both plates have a constant thickness. Express your result in terms of the total system mass.

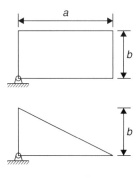

Figure 1.4

CHAPTER 2

SINGLE-DEGREE-OF-FREEDOM SYSTEMS

As mentioned in Chapter 1, in many cases an approximate analysis involving only a limited number of degrees of freedom will provide sufficient accuracy for evaluating the dynamic response of a structural system. The single-degree-of-freedom (SDOF) system will represent the simplest solution to the dynamic problem. Therefore, initial consideration will be given to a system having only a single degree of freedom that defines the motion of all components of the system. For many systems, this will depend on the assemblage of the members and the location of external supports or internal hinges. In order to develop an SDOF model of the actual structure, it is necessary to reduce the continuous system to an equivalent discrete system having a displaced shape that is defined in terms of a single displacement coordinate. The resulting representation of the actual structure is often referred to as the *discretized model*.

2.1 REDUCTION OF DEGREES OF FREEDOM

In order to reduce the degrees of freedom, one of the following methods is often used:

1. Lumped parameters
2. Assumed deflection pattern
3. A deflection pattern based on the static deflected shape

A common practice is to lump the participating mass of the structure at one or more discrete locations. For example, in the case of a simply supported beam, the participating mass may be located at the center of the simple span with the stiffness represented by a weightless beam with distributed elasticity. Alternatively, the same beam may have the elasticity represented by a concentrated resistance (spring) at the center of the span connecting two rigid segments having distributed mass.

For many simple dynamic systems, it may be possible to accurately estimate the displaced shape. For example, the displacement at each point along the beam may be estimated as

$$v(x,t) = \sin \frac{\pi x}{L} y(t) = \phi(x)y(t) \tag{2.1}$$

where L = the length of the span
x = the distance from the left support
$y(t)$ = the displacement at the center of the span

The term $\sin(\pi x/L)$ may be considered as a displacement shape function, which depends on location on the structure. For the simply supported beam, the sine function gives a close approximation to the actual deformed shape.

For more complex structures, such a simple shape function may be difficult to obtain. A third procedure that was suggested by Lord Rayleigh uses the static deflected shape of the structural system as the shape function. This provides a powerful method for transforming a complex system with multiple degrees of freedom into an equivalent SDOF system. It requires that a static representation of the dynamic loads be applied to the structural system and the resulting normalized deflection pattern be used as the shape function. In the case of the previous simply supported beam, the displacement can be represented as

$$v(x,t) = \phi(x)y(t) \tag{2.2}$$

where

$$\phi(x) = \frac{v(x)}{v_{max}} \tag{2.3}$$

is the shape function, which is the static displacement normalized by the maximum static displacement that occurs at the center of the beam.

2.2   TIME-DEPENDENT FORCE

The simplest structure that can be considered for dynamic analysis is an idealized one-story structure in which the single degree of freedom is the lateral translation at the roof level, as shown in Figure 2.1a. In this idealization, three important assumptions are made. First, the mass is assumed to be concentrated (lumped) at the roof level. Second, the roof system is assumed to be rigid, and third, the axial deformation in the columns is neglected. From these assumptions, it follows that all lateral resistance is provided by resisting elements such as columns, walls, or diagonal braces located between the roof and the base. Application of these assumptions results in a discretized structure that can be represented as shown in either Figure 2.1b or Figure 2.1c with a time-dependent force applied at the roof level. The total stiffness, k, is simply the sum of the stiffness of each resisting element in the story level.

The forces acting on the mass of the structure are shown in Figure 2.1d. Summing the forces acting on the free body results in the following equation of equilibrium, which must be satisfied at each instant of time:

$$f_i + f_d + f_s = p(t) \qquad (2.4)$$

where f_i = the inertia force = $m\ddot{u}$
  f_d = the damping (dissipative) force = $c\dot{v}$
  f_s = the elastic restoring force = kv
  $p(t)$ = the time-dependent applied force
  $\ddot{u}$ = the total acceleration of the mass
  $\dot{v}, v$ = the velocity and displacement of the mass relative to the
    base

Writing Equation (2.4) in terms of the physical response parameters results in

$$m\ddot{u} + c\dot{v} + kv = p(t) \qquad (2.5)$$

It should be noted that the forces in the damping element and in the resisting elements depend on the relative velocity and relative displacement, respectively, across the ends of these elements, whereas the inertia force depends on the total acceleration of the mass. The total acceleration of the mass can be expressed as

$$\ddot{u}(t) = \ddot{v}_g(t) + \ddot{v}(t) \qquad (2.6)$$

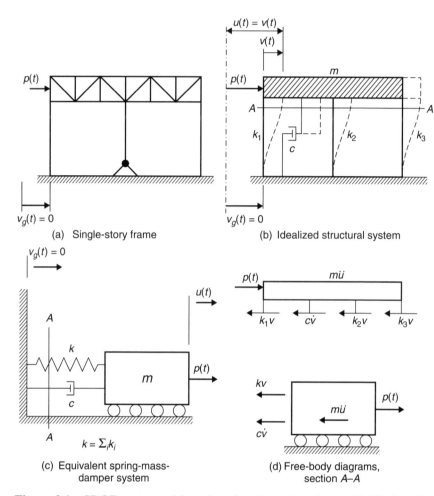

(a) Single-story frame

(b) Idealized structural system

(c) Equivalent spring-mass-
damper system

(d) Free-body diagrams,
section A–A

Figure 2.1 SDOF system subjected to time-dependent forces (F. Naeim, *The Seismic Design Handbook*, 2nd ed. (Dordrecht, Netherlands: Springer, 2001), reproduced with kind permission from Springer Science + Business Media B.V.)

where $\ddot{v}(t)$ = the acceleration of the mass relative to the base

$\ddot{v}_g(t)$ = the acceleration of the base

Substituting Equation (2.6) into Equation (2.5) results in an alternate form of Equation (2.5) with zero-applied force:

$$m\ddot{v} + c\dot{v} + kv = p_e(t) = -m\ddot{v}_g$$

where $p_e(t)$ is referred to as the *effective dynamic force*.

In the case of a time-dependent dynamic force, the base is assumed to be fixed with no motion, and hence $\ddot{v}_g(t) = 0$ and $\ddot{u}(t) = \ddot{v}(t)$. Making this substitution for the acceleration, Equation (2.5) for a time-dependent force becomes

$$m\ddot{v} + c\dot{v} + kv = p(t) \tag{2.7}$$

2.3 GRAVITATIONAL FORCES

Consider a simply supported beam with half of the mass lumped at the center of the span and a quarter of the mass lumped over each support. In this case, the force of gravity acts in the direction of the displacement, and the mass over the supports does not participate in the response. For the system of forces acting on the mass, the condition of equilibrium can be written as

$$m\ddot{v} + c\dot{v} + kv = p(t) + W \tag{2.8}$$

The total displacement, v, is expressed as the sum of the static displacement, v_s, caused by the weight and the additional displacement, v_d, due to the dynamic load, $p(t)$:

$$v = v_s + v_d \tag{2.9}$$

The resisting force in the spring can be expressed as

$$f_s = kv = kv_s + kv_d \tag{2.10}$$

Introducing this expression into Equation (2.8) results in the following expression for dynamic equilibrium:

$$m\ddot{v} + c\dot{v} + kv_s + kv_d = p(t) + W \tag{2.11}$$

Because $kv_s = W$, and $\dot{v}_s, \ddot{v}_s$ do not vary with time, Equation (2.11) results in the following equation:

$$m\ddot{v}_d + c\dot{v}_d + kv_d = p(t) \tag{2.12}$$

This indicates that the equation of motion that was developed with reference to the static equilibrium position of the dynamic forces is not affected by the gravity forces. For this reason, the equation of motion will be referenced from the static position. However, the total displacements and stresses must be obtained by adding the static quantities to the results of the dynamic analysis.

2.4 EARTHQUAKE GROUND MOTION

When a single-story structure, as shown in Figure 2.2a, is subjected to earthquake ground motions, no external dynamic force is applied at the roof level. Instead, the system experiences an acceleration of the base. The effect of this on the idealized structure is shown in Figure 2.2b

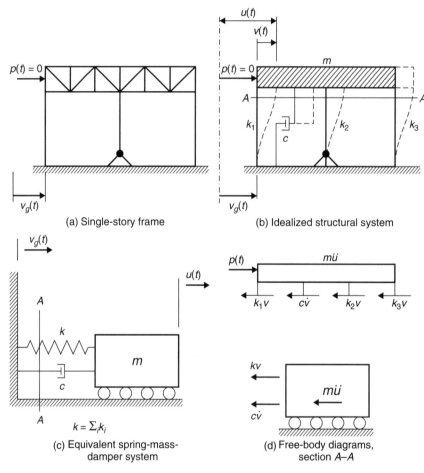

(a) Single-story frame

(b) Idealized structural system

(c) Equivalent spring-mass-damper system

(d) Free-body diagrams, section A–A

Figure 2.2   SDOF system subjected to base motion (F. Naeim, *The Seismic Design Handbook*, 2nd ed. (Dordrecht, Netherlands: Springer, 2001), reproduced with kind permission from Springer Science + Business Media B.V.)

and 2.2c. Summing the forces shown in Figure 2.2c results in the following equation of dynamic equilibrium:

$$m\ddot{u} + c\dot{v} + kv = 0 \tag{2.13}$$

This equation can be written in the form of Equation (2.7) by substituting Equation (2.6) into Equation (2.13) and rearranging terms to obtain

$$m\ddot{v} + c\dot{v} + kv = p_e(t) \tag{2.14}$$

where $p_e(t) = $ the effective time-dependent force $= -m\ddot{v}_g(t)$

Hence, the equation of motion for a structure subjected to a base motion is similar to that for a structure subjected to a time-dependent force if the effect of the base motion is represented as an effective time-dependent inertia force that is equal to the product of the mass and the ground acceleration.

2.5 FORMULATION OF EQUATION OF MOTION

2.5.1 d'Alembert's Principle

The concept of d'Alembert's principle has already been introduced. Application of this principle is very convenient to use in certain problems in structural dynamics because it allows the equation of motion to be reduced to an equivalent equation of static equilibrium. In many simple problems, the most direct and convenient way of formulating the equation of motion is by this procedure.

Example 2.1   Use d'Alembert's principle to develop the equation of motion (dynamic equilibrium) for the rigid, weightless bar shown in Figure 2.3. Note that the mass of the bar has been lumped at two locations, the stiffness of the system has been represented as a concentrated spring, and a damping element has been included.

$$\sum M_A = 0$$

$$m(0.75a)^2\ddot{\theta} + ma^2\ddot{\theta} + ca^2\dot{\theta} + ma^2\ddot{\theta} - F(t)1.6a + (2.2a)^2k\theta = 0$$

$$2.56ma\ddot{\theta} + ca\dot{\theta} + 4.84ak\theta = 1.6F(t)$$

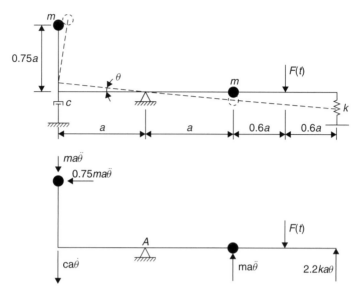

Figure 2.3 d'Alembert's principle, discrete parameters

2.5.2 Virtual Work (Virtual Displacements)

By applying the principle of virtual work in terms of virtual displacements to a spring-mass-damper system, subjected to a driving force $f(t)$, we can write the equation of virtual work as

$$[f(t) - m\ddot{v} - c\dot{v} - kv]\, \delta v = 0 \tag{2.15}$$

Because the virtual displacement cannot be zero, the term in brackets must be zero to satisfy the work equation, and therefore the equation of motion (dynamic equilibrium) becomes

$$m\ddot{v} + c\dot{v} + kv = f(t) \tag{2.16}$$

Example 2.2 Use the principle of virtual displacements to develop the equation of motion for the SDOF system shown in Figure 2.4. The bar is rigid and has a total mass, m.

$$p(x,t) = \frac{x}{L}p_0(t)$$

$$v(x,t) = x \sin\theta(t)$$

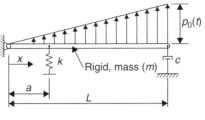

Figure 2.4

For small $\theta(t)$, $\sin \theta(t) \cong \theta(t)$

$$\therefore v = x\theta \quad \text{and} \quad \delta v = x\delta\theta$$

The equation of dynamic equilibrium can be written as

$$-M_i - f_s a - f_d L + \frac{2Lf_p}{3} = 0$$

Multiply by the virtual displacement, $\delta\theta$, to obtain the virtual work equation (see Figure 2.5).

$$\delta W = 0 = -M_i \delta\theta - f_s a\delta\theta - f_d L\delta\theta + f_p \left(\frac{2L}{3}\right)\delta\theta$$

where $M_i = \dfrac{mL^2}{3}\ddot{\theta}$, $\quad f_s = ka\theta$, $\quad f_d = cL\dot{\theta}$, and $f_p = \dfrac{L}{2}p_0(t)$.

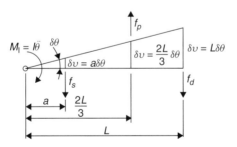

Figure 2.5   Application of virtual displacements

Substituting into the work equation:

$$\left(-\frac{mL^2}{3}\ddot{\theta} - cL^2\dot{\theta} - ka^2\theta + \frac{p_0 L^2}{3}\right)\delta\theta = 0$$

Because $\delta\theta \neq 0$, the quantity in parentheses is equal to zero.

The equation of motion becomes

$$\frac{mL^2}{3}\ddot{\theta} + cL^2\dot{\theta} + ka^2\theta = \frac{L^2}{3}p_0(t)$$

2.6 GENERALIZED COORDINATES

As indicated in Equation (2.2), the generalized expression for the displacements in an SDOF system can be written as

$$v(x,t) = \phi(x)y(t) \tag{2.17}$$

This is an application of the classical mathematical technique of separation of variables, except that, in this case, the shape function, $\phi(x)$, is declared a priori instead of being derived mathematically.

For any assumed displacement function, $\phi(x)$, the resulting shape depends on the time-dependent amplitude, $y(t)$, which is referred to as the *generalized coordinate*. The spatial shape function, $\phi(x)$, relates the structural degrees of freedom to the generalized coordinate. For most structural systems, it is necessary to represent the restoring forces in the damping elements and the stiffness elements in terms of the relative velocity and relative displacement between the ends of the element, respectively:

$$\Delta\dot{v}(x,t) = \Delta\phi(x)\dot{y}(t) \tag{2.18}$$

$$\Delta v(x,t) = \Delta\phi(x)y(t) \tag{2.19}$$

Once the displacement function is selected, the structure is constrained to deform in that prescribed manner. This implies that the displacement function must be selected carefully if a good approximation of the dynamic properties and response of the system are to be obtained using this simplified model. This section will develop the equations for determining the generalized response parameters in terms of the spatial displacement function and the physical member properties. Methods for determining the shape function will be discussed, and techniques for determining the more correct displacement function for a particular structure will be presented. A useful procedure known as *Rayleigh's method*, which uses the static deflected shape of the structure, will be introduced for discrete parameters.

2.6.1   Discrete Parameters

Initially, the generalized properties will be developed for systems that consist of an assemblage of lumped masses, discrete damping elements, and discrete structural elements. For a time-dependent force, the condition of dynamic equilibrium is given by Equation (2.7). Applying the principle of virtual work in the form of virtual displacements results in an equation of virtual work in the form

$$f_i \delta v + f_d \delta \Delta v + f_s \delta \Delta v - p(t)\delta v = 0 \qquad (2.20)$$

where it is understood that $v = v(x, t)$ and that the virtual displacements applied to the damping force and the elastic restoring force are virtual, relative displacements and that the inertia force may be caused by translational motion, rotational motion, or a combination of both. These virtual displacements can be expressed as

$$\delta v(x, t) = \phi(x)\delta y(t) \qquad (2.21)$$

and the virtual relative displacement can be written as

$$\delta \Delta v(x, t) = \Delta \phi(x)\delta y(t) \qquad (2.22)$$

where

$$\Delta v(x, t) = \phi(x_i)y(t) - \phi(x_j)y(t) = \Delta \phi(x)y(t)$$

The inertia, damping, and elastic restoring forces can be expressed as

$$f_i = m\ddot{v} = m\phi\ddot{y}$$
$$f_d = c\,\Delta \dot{v} = c\,\Delta \phi \dot{y} \qquad (2.23)$$
$$f_s = k\,\Delta v = k\,\Delta \phi y$$

Substituting Equations (2.21), (2.22), and (2.23) into Equation (2.20) results in the following equation of virtual work in terms of δ_y:

$$m\phi\ddot{y}\phi\delta_y + c\,\Delta \phi \dot{y}\,\Delta \phi \delta_y + k\,\Delta \phi y\,\Delta \phi \delta_y = p_i \phi_i \delta_y \qquad (2.24)$$

Factoring δy:

$$(m\phi\ddot{y}\phi + c\,\Delta \phi \dot{y}\,\Delta \phi + k\,\Delta \phi y\,\Delta \phi - p\phi)\delta y = 0$$

and recognizing that δy cannot be zero, we can express the following equation of motion in terms of the generalized coordinate and generalized parameters:

$$m^*\ddot{y} + c^*\dot{y} + k^*y = p^*(t) \tag{2.25}$$

where m^*, c^*, k^*, and p^* are referred to as the *generalized parameters*, which are defined as

$$m^* = \sum_i m_i \phi_i^2 = \text{generalized mass}$$

$$c^* = \sum_i c_i \Delta\phi_i^2 = \text{generalized damping}$$

$$k^* = \sum_i k_i \Delta\phi_i^2 = \text{generalized stiffness} \tag{2.26}$$

$$p^* = \sum_i p_i \phi_i = \text{generalized force}$$

For a time-dependent base acceleration, the generalized force becomes

$$p^* = \ddot{v}_g(t) \sum_i m_i \phi_i \tag{2.27}$$

The effect of the generalized-coordinate approach is to transform a multiple-degree-of-freedom (MDOF) dynamic system into an equivalent SDOF system in terms of the generalized coordinate. This transformation is shown schematically in Figure 2.6. The degree to which the response of the transformed system represents the actual system will depend on how well the assumed displacement shape represents the dynamic displacement of the actual structure.

For building structures, the displacement shape depends on the aspect ratio of the structure, which is defined as the ratio of the height to the base dimension. Possible shape functions for high-rise, midrise, and low-rise structures are summarized in Figure 2.7. The stiffness properties of some common building elements are summarized in Figure 2.8.

Once the dynamic response is obtained in terms of the generalized coordinate, Equation (2.17) must be used to determine the displacements in the structure, and these, in turn, can be used to determine the forces in the individual structural elements.

In principle, any function that represents the general deflection characteristics of the structure and satisfies the support conditions can be

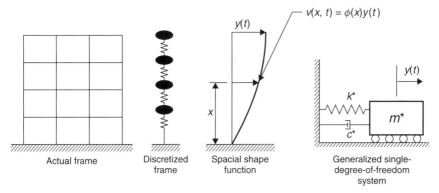

Figure 2.6 Generalized SDOF discretization (F. Naeim, *The Seismic Design Handbook*, 2nd ed. (Dordrecht, Netherlands: Springer, 2001), reproduced with kind permission from Springer Science + Business Media B.V.)

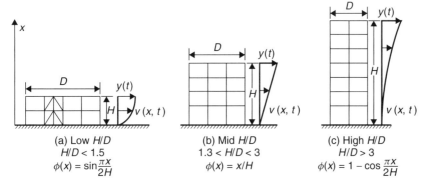

Figure 2.7 Possible deflected shapes based on aspect ratio (F. Naeim, *The Seismic Design Handbook*, 2nd ed. (Dordrecht, Netherlands: Springer, 2001), reproduced with kind permission from Springer Science + Business Media B.V.)

used. However, any shape other than the true vibration shape requires the addition of external constraints to maintain equilibrium. These extra constraints tend to stiffen the system and thereby increase the computed frequency. The true vibration shape will have no external constraints and therefore will have the lowest frequency of vibration. When choosing between several approximate deflected shapes, the one producing the lowest frequency is always the best approximation. A good approximation of the true vibration shape can be obtained by applying forces representing the inertia forces and letting the static deformation of the structure determine the spatial shape function.

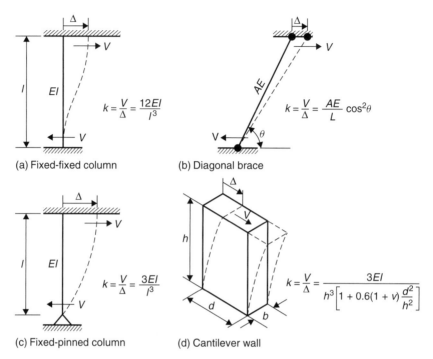

(a) Fixed-fixed column   (b) Diagonal brace

(c) Fixed-pinned column   (d) Cantilever wall

Figure 2.8   Stiffness properties of common lateral force–resisting elements

Example 2.3   The motion of the rigid-body system shown in Figure 2.9 can be expressed in terms of a single degree of freedom represented by the angle of rotation about the pinned support. Use this degree of freedom as the generalized coordinate and develop the equation of motion for the system.

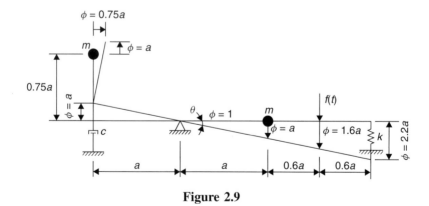

Figure 2.9

The equation of motion has the general form

$$m^*\ddot{\theta} + c^*\dot{\theta} + k^*\theta - p^*(t) = 0$$

The generalized parameters are determined as

$$m^* = \sum m_i \phi_i^2 = m(a)^2 + m(a)^2 + m(0.75a)^2 = 2.56ma^2$$

$$c^* = \sum c_i (\Delta\phi_i)^2 = c(a)^2$$

$$k^* = \sum k_i (\Delta\phi_i)^2 = k(2.2a)^2 = 4.84a^2$$

$$p^*(t) = \sum p_i \phi_i = 1.6f(t)a$$

Substituting for the generalized parameters and dividing through by a, we get

$$2.56ma\ddot{\theta} + ca\dot{\theta} + 4.84ka\theta = 1.6f(t)$$

Example 2.4   Using the displacement coordinate, θ, as the generalized coordinate, develop the equation of motion in terms of this displacement, as shown in Figure 2.10.

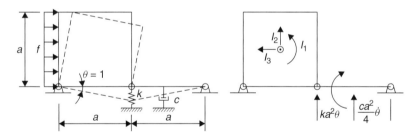

Figure 2.10

For the plate, $m_p = m_0$ (per unit area) a^2; and for the beam, $m_b = m_1$ (per unit length)a.

$$\text{For the plate} : m_p^* = m_0 \left(\frac{a^2}{6} + \frac{a^2}{2} \right) a^2 = \frac{2}{3}m_0 a^4$$

$$\text{For the beam} : m_b^* = m_1 \frac{a^2}{3}$$

$$\text{For the damper} : c^* = c \left(\frac{a}{2} \right)^2 = \frac{ca^2}{4}$$

For resistance : $k^* = ka^2$

For loading : $f^* = fa\dfrac{a}{2} = \dfrac{fa^2}{2}$

Dividing terms by a^2 and combining terms, we can rewrite the equation of motion as

$$\left(\frac{2m_0 a^2}{3} + \frac{m_1 a}{3}\right)\ddot{\theta} + \frac{c}{4}\dot{\theta} + k\theta = \frac{f}{2}$$

Example 2.5   The four-story steel frame shown in Figure 2.11 is idealized as a space frame with moment connections in both directions. The two following assumptions are used to idealize the mathematical model: (a) the girders are assumed to be rigid relative to the columns, and (b) the floor diaphragm is assumed to be rigid in its own plane. The live load for the roof is 0.020 kip/ft^2 and for the floors, 0.050 kip/ft^2. It should be noted that, in general, the live load is not combined with the seismic load but is given as a reference for the design process. The occupancy is assumed to be office space. The total dead load (self-weight) for the four floor levels is estimated below. The total weight of the second level is 2 kips more than that of a typical floor because of the increased story height. Determine the generalized mass and stiffness for an SDOF model.

Roof Level

Penthouse	10 kips
Floor slab ($0.110 \times 4.5/12 \times 40 \times 80$)	131 kips
Ceiling ($0.010 \times 40 \times 80$)	32 kips
Roofing ($0.005 \times 40 \times 80$)	16 kips
Exterior precast concrete panels	44 kips
Steel framing	45 kips
Miscellaneous	16 kips
Total	294 kips

Typical Floor Level

Floor slab ($0.110 \times 4.5/12 \times 40 \times 80$)	131 kips
Ceiling ($0.010 \times 40 \times 80$)	32 kips
Fixed partitions ($0.020 \times 40 \times 80$)	64 kips
Exterior precast concrete panels	68 kips

Steel framing	57 kips
Miscellaneous	16 kips
Total	368 kips

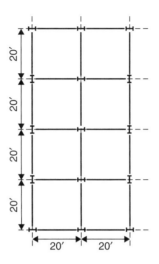

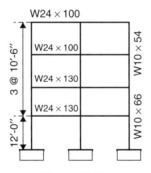

Figure 2.11

Using the two assumptions mentioned previously, we can calculate the lateral stiffness of the columns as $k = 12EI/L^3$, which has the following values for the different directions and different floor levels:

Third and fourth stories: W10 × 54: $I_{xx} = 306$ in^4, $I_{yy} = 103.9$ in^4, $L = 10.5$ ft

$$k_{xx} = \frac{12(29,000)306}{[(10.5)(12)]^3} = 53.23\frac{\text{kips}}{\text{in}} \quad k_{yy} = \frac{12(29,000)104}{[(10.5)(12)]^3} = 18.1\frac{\text{kips}}{\text{in}}$$

Second story: W10 × 66: $I_{xx} = 382.5\,\text{in}^4, I_{yy} = 129.2\,\text{in}^4, L = 10.5\,\text{ft}$

$$k_{xx} = \frac{12(29,000)382.5}{[(10.5)(12)]^3} = 66.54\frac{\text{kips}}{\text{in}} \quad k_{yy} = \frac{12(29,000)129.2}{[(10.5)(12)]^3} = 22.48\frac{\text{kips}}{\text{in}}$$

First story: W10 × 66: $I_{xx} = 382.5\,\text{in}^4, I_{yy} = 129.2\,\text{in}^4, L = 12.0\,\text{ft}$

$$k_{xx} = \frac{12(29,000)382.5}{[(12)(12)]^3} = 44.58\frac{\text{kips}}{\text{in}} \quad k_{yy} = \frac{12(29,000)129.2}{[(12)(12)]^3} = 15.06\frac{\text{kips}}{\text{in}}$$

The story stiffness in the transverse direction can then be calculated as

$$K_i = \sum k_{xx} + \sum k_{yy}$$

Fourth and third stories: $\quad K = 9(53.23) + 6(18.1) = 587.7 \text{ kips/in}$
Second story: $\quad K = 9(66.54) + 6(18.1) = 733.7 \text{ kips/in}$
First story: $\quad K = 9(44.58) + 6(15.1) = 491.6 \text{ kips/in}$

Because the aspect ratio for this frame is

$$\frac{H}{D} = \frac{3(10.5) + 12}{40} = 1.09 \le 1.5$$

it would be reasonable to assume the deflected shape function can be represented by

$$\phi(x) = \sin\frac{\pi x}{2L}$$

as shown in Figure 2.7.

Level	K	M	ϕ_i	$\Delta\phi_i$	$M_i\phi_i^2$	$K_i(\Delta\phi_i)^2$
4		0.761	1.000		0.761	
	587.7			0.071		2.96
3		0.952	0.929		0.822	
	587.7			0.203		24.22
2		0.952	0.726		0.502	
	733.7			0.306		68.70
1		0.958	0.420		0.169	
	491.6			0.420		86.72

From these values, the generalized properties can be determined as

$$m^* = \sum m_i \phi_i^2 = 2.25 \text{ kip-sec}^2/\text{in}$$

and

$$k^* = \sum k_i (\Delta\phi_i)^2 = 182.6 \text{ kips/in}$$

2.6.2 Continuous Parameters

A system with continuous parameters is one in which the mass, stiffness, and force may vary with the position, x, in the structure. Hence, the distributed mass of the system is represented in terms of the mass per unit length, $\mu(x)$, and the time-dependent force may vary with both position and time, $f(x,t)$. For convenience, the damping is represented as a distributed damping, $c(x)$. Therefore, the inertia force and the damping force have the following form, respectively:

$$f_i = \mu(x)\phi(x)\ddot{y}(t) \tag{2.28}$$

$$f_d = c(x)\phi(x)\dot{y}(t) \tag{2.29}$$

In most cases, the deformations are a result of flexure, which is proportional to the curvature; however, they also could be caused by either shear or axial force. For the case of flexure, the flexural forces can be expressed as

$$f_s = EI(x)v''(x,t) = EI(x)\phi''(x)y(t) \tag{2.30}$$

The virtual displacement is $\delta v = \phi(x)\delta y(t)$, and the virtual curvature becomes

$$\delta v''(x,t) = \phi''(x)\delta y(t) \tag{2.31}$$

By making these substitutions, the equation of virtual work becomes

$$\int_0^L [f_i + f_d]\delta v\, dx + \int_0^L f_s \delta v''\, dx - \int_0^L p(x,t)\delta v\, dx = 0$$

On substitution for the virtual displacement, noting that $\delta v \neq 0$, the equation of motion becomes

$$m^*\ddot{y}(t) + c^*\dot{y}(t) + k^* y(t) = p^*(t) \tag{2.32}$$

where

$$m^* = \int_0^L \mu(x)\phi^2(x)dx \qquad c^* = \int_0^L c(x)\phi^2(x)dx$$

$$k^* = \int_0^L EI(x)\left[\phi''(x)\right]^2 dx \qquad p^* = \int_0^L p(x,t)\phi(x)dx$$

In a similar manner, the restoring forces for displacements in the axial direction can be determined as

$$f_s = AE(x)u'(x,t) = \sigma_x A(x)$$

where u = axial deformation

$$\sigma_x = Eu'$$
$$u'(x,t) = \phi'(x)y(t)$$
$$\delta u'(x,t) = \phi'(x)\delta y(t)$$
$$f_s = AE(x)\phi'(x)y(t)$$

Substituting for the restoring force in the work equation results in the following term for the generalized stiffness:

$$k^* = \int_0^L AE(x)\left[\phi'(x)\right]^2 dx$$

Example 2.6 Determine the generalized stiffness, generalized mass, and generalized force for the simply supported beam with uniform load shown in Figure 2.12, using the following two shape functions:

a. $\phi_1(x) = \sin\dfrac{\pi x}{L}$

b. $\phi_2(x) = 3.2\left[\dfrac{x}{L} - \dfrac{2x^3}{L^3} + \dfrac{x^4}{L^4}\right]$

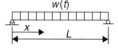

Figure 2.12

For the shape function, ϕ_1, the generalized stiffness can be written as

$$k^* = \int_0^L EI\left[\phi_1''(x)\right]^2 dx = EI\int_0^L \left[\left(-\frac{\pi}{L}\right)^2 \sin\frac{\pi x}{L}\right]^2 dx = 48.7\frac{EI}{L^3}$$

For the polynomial shape function, ϕ_2, which represents the static deflected shape, the generalized stiffness is

$$k^* = \int_0^L EI\left[\phi_2''(x)\right]^2 dx = (38.4)^2 EI \int_0^L \left(\frac{x^2}{L^6} - 2\frac{x^3}{L^7} + \frac{x^4}{L^8}\right) dx = \frac{49.15EI}{L^3}$$

These results indicate that both of these shape functions give a good estimate of the actual stiffness. Because the true stiffness is the one with no artificial constraints, the lower stiffness provides the best estimate.

Using these two shape functions, we can estimate the generalized mass for the simply supported beam as

$$m^* = \int_0^L \mu(x)\phi^2(x)dx = \mu \int_0^L \sin^2 \frac{\pi x}{L}dx = \frac{\mu L}{2} = \frac{M}{2}$$

where M is the total mass of the system. For the polynomial representation of the deflected shape,

$$m^* = \int_0^L \mu(x)\phi^2(x)dx = \mu(3.2)^2 \int_0^L \left(\frac{x}{L} - \frac{2x^3}{L^3} + \frac{x^4}{L^4}\right)^2 dx = 0.502M$$

Similarly, the generalized force based on the two shape functions can be estimated as

$$p^* = \int_0^L w(x,t)\phi(x)dx = w \int_0^L \sin \frac{\pi x}{L}dx = 0.64W$$

where $W = wL$ is the total load on the member:

$$p^* = \int_0^L w(x,t)\phi(x)dx = 3.2w \int_0^L \left(\frac{x}{L} - \frac{2x^3}{L^3} + \frac{x^4}{L^4}\right)dx = 0.64W$$

Example 2.7(M)   Solve Example 2.6 using the MATLAB Symbolic Math Toolbox.

MATLAB's Symbolic Math Toolbox provides a very powerful tool for carrying out symbolic integrations and differentiations such as the ones presented in Example 2.5. Because this is the first example using MATLAB in this book, we will take the time to elaborate on every step of the use of MATLAB to solve this problem.

It is good practice to clear all variables created previously and to clear the MATLAB command window. This way, we will see the results of what we are doing at the moment and will not be confused by the results of previous calculations. You can do this using the functions `clear all` and `clc`. If you have other MATLAB windows open for figures and other things, you can use the `close all` function to close them.

Next, we have to define our symbolic variables, which are E, I, x, L, w, and μ. Because MATLAB does not allow definition of variables as Greek symbols, we use the variable mu instead of μ. We use the `syms` statement to declare our symbolic variables:

```
syms E I x L mu w
```

Likewise, $\phi_1(x)$ and $\phi_2(x)$ can be represented by the MATLAB variables `phi1` and `phi2`:

```
phi1=sin(pi*x/L);
phi2=3.2*(x/L-2*x^3/L^3+x^4/L^4);
```

The semicolon (;) is added to the end to prevent the echo of immediate results on the screen. We perform differentiation using the `diff` function in the form of `diff(f, x, n)`, where f is the function to be operated on, x is the variable with respect to which the differentiation is to be carried out, and n is the number of differentiations to be performed. Therefore, $\phi_1''(x)$ and $\phi_2''(x)$ can be defined as

```
phi1pp=diff(sin(pi*x/L),x,2);
phi2pp=diff(phi2,x,2);
```

Finally, we perform integration using the `int` function in the form of `int(f, x, a, b)`, where f is the function to be operated on, x is the variable with respect to which the integration is to be carried out, and a and b are the lower and upper limits of integration. If a and b are not specified, an indefinite integral of f is carried out with respect to x. Therefore, for shape functions $\phi_1(x)$ and $\phi_2(x)$, k^* values are obtained from

```
K1=int(E*I*phi1pp^2,x,0,L);
K2=int(E*I*phi2pp^2,x,0,L);
```

Similarly, the values for m^* and p^* are obtained from

```
M1=int(mu*phi1^2,x,0,L);
M2=int(mu*phi2^2,x,0,L);
P1=int(w*phi1,x,0,L);
P2=int(w*phi2,x,0,L);
```

We can use the vpa function to simplify fractions and present results in decimal form with a specified number of digit accuracy. For example, K1=vpa(K1,3) results in presentation of the expression calculated for K1 using three significant digits.

The entire MATLAB script for this example is as follows (% denotes comments):

```
clear all
clc
syms E I x L mu w
% solving for the first shape function
phi1=sin(pi*x/L);              % this is the first shape function
phi1pp=diff(sin(pi*x/L),x,2);  % this is the second derivative
                               % of the shape function
K1=int(E*I*phi1pp^2,x,0,L);    % integrating
K1=vpa(K1,3)      % displaying numeric value with three digits
                  % after decimal
M1=int(mu*phi1^2,x,0,L);
M1=vpa(M1,3)
P1=int(w*phi1,x,0,L);
P1=vpa(P1,3)
% solving for the second shape function
phi2=3.2*(x/L-2*x^3/L^3+x^4/L^4);  % this is the second shape
                                   % function
phi2pp=diff(phi2,x,2); % this is the second derivative of the
                       % shape function
K2=int(E*I*phi2pp^2,x,0,L); % integrating
K2=vpa(K2,3) % displaying numeric value with three digits after
             % decimal
M2=int(mu*phi2^2,x,0,L);
M2=vpa(M2,3)
P2=int(w*phi2,x,0,L);
P2=vpa(P2,3)
```

Upon execution, MATLAB will display the following results. Note that MATLAB displays the variable name and the results on subsequent lines. In this book in order to save space, we show both on a single line.

```
K1 = (48.7*E*I)/L^3
M1 = 0.5*L*mu
P1 = 0.637*L*w
K2 = (49.2*E*I)/L^3
M2 = 0.504*L*mu
P2 = 0.64*L*w
```

The generalized SDOF concepts that have been discussed can also be applied to reduce a two-dimensional system in the horizontal plane into an equivalent SDOF system. This procedure will be illustrated in the following example.

Example 2.8   A square uniform slab having length a on each side is simply supported on all sides. If the mass per unit area is γ, the external loading per unit area is $\bar{p}(t)$ and the flexural rigidity is

$$D = \frac{Eh^3}{12(1 - v^2)}$$

Determine the equation of motion in terms of the generalized displacement, $Y(t)$, at the center of the plate. Assume the shape function for the displaced shape is

$$\psi(x, y) = \sin\frac{\pi x}{a}\sin\frac{\pi y}{a}$$

The generalized mass can be written as

$$m^* = \int_A m(x, y)\left[\psi(x, y)\right]^2 dA \quad m^* = \gamma \int_0^a \int_0^a \sin^2\frac{\pi x}{a}\sin^2\frac{\pi y}{a}dxdy$$

Simplifying this expression results in the generalized mass given by

$$m^* = \frac{\gamma a^2}{4}$$

The generalized stiffness can be written as

$$k^* = D\int_A\left[\left(\frac{\partial^2\psi}{\partial x^2} + \frac{\partial^2\psi}{\partial y^2}\right)^2 - 2(1 - v)\left(\frac{\partial^2\psi}{\partial x^2}\frac{\partial^2\psi}{\partial y^2} - \frac{\partial^2\psi}{\partial x\partial y}\right)\right]dA$$

Performing the double integrations and simplifying:

$$k^* = D\left[4\left(\frac{\pi}{a}\right)^4\left(\frac{a}{2}\right)^2 - 2(1 - v)\left(\frac{\pi}{a}\right)^4\left(\frac{a}{2}\right)^2 + 2(1 - v)(0)\right]$$

$$k^* = \frac{D\pi^4}{a^2}\left[1 - \frac{1}{2}(1 - v)\right] = \frac{D\pi^4}{2a^2}(1 + v)$$

and the generalized loading is

$$p^*(t) = \int_A p(x, y)\psi(x, y)dA$$

$$p^*(t) = \bar{p}\int_0^a \int_0^a \sin\frac{\pi x}{a}\sin\frac{\pi y}{a}dydx = \frac{4\bar{p}a^2}{\pi^2} = 0.41\bar{p}a^2$$

Example 2.9(M)   Solve Example 2.8 using the MATLAB Symbolic Math Toolbox.

The only difference in integration for this problem compared to Example 2.7(M) is that this problem involves double integration. Double integration can be carried out in two steps (e.g., once with respect to y and then with respect to x), or it can be performed in one step by calling an `int` function from within an `int` function. An example of integration in two steps is as follows:

```
% generalized mass
   yint=int(phi^2,y,0,a);     % integrate with respect to y
   M=mu*int(yint,x,0,a)       % now integrate with respect to x
```

We could have done the same thing in a single step:

```
% generalized mass
   M=mu*int(int(phi^2,y,0,a),x,0,a); % double integration
                                     % in one step
```

The entire MATLAB script for this example is as follows:

```
clear all
clc
syms E h mu x y a D nu
% Define D
D=(E*h^3)/(12.*(1-nu^2));
% define shape function
   phi=sin(pi*x/a)*sin(pi*y/a);  % shape function
% generalized mass
   yint=int(phi^2,y,0,a);        % integrate with respect to y
   M=mu*int(yint,x,0,a)          % now integrate with respect to x
% generalized loading
   P=mu*int(int(phi,y,0,a),x,0,a);% double integration in one step
   P=vpa(P,2)                     % simplify to show three digits
% Generalized Stiffness
   phippx=diff(phi,x,2);
   phippy=diff(phi,y,2);
   phippxy=diff(diff(phi,y,1),x,1);
   K=D*int(int((phippx+phippy)^2 -2*(1-nu)
            *(phippx*phippy-phippxy),y,0,a),x,0,a)
```

Upon execution, MATLAB will display the following results:

```
M = (a^2*mu)/4
P = 0.41*a^2*mu
K = -(E*pi^4*h^3*(nu + 1))/(2*a^2*(12*nu^2 - 12))
```

2.6.3 Transformation Factors

The properties for a generalized SDOF system can also be obtained by use of tabulated transformation factors that are obtained using the procedures just discussed. This is done by multiplying the total load, mass, resistance, and stiffness of the real structure by the corresponding transformation factor. The mass factor, K_M, is defined as the ratio of the mass of the equivalent system, m^*, to the total mass of the system, M, and can be expressed as

$$K_M = \frac{m^*}{M}$$

which, for the simply supported beam, results in $K_M = 0.5$. The load factor is defined as the ratio of the load on the equivalent system, p^*, to the total applied load, F, as

$$K_L = \frac{p^*}{F}$$

which, for the simply supported beam, results in $K_L = 0.64$. It should be noted that this definition of load factor applies to the magnitude of the force, and both the equivalent load and the real load have the same time function. The resistance of an actual member or structure can vary depending on the material properties. The resistance is the internal force tending to restore the unloaded element (structure) to its unloaded static position. For the purposes of this analysis, the resistance functions must be simplified; therefore, it will be assumed that the element (structure) has an elastoplastic resistance. In this case, the maximum resistance, R_m, is the plastic limit load that the member can support statically, and the resistance factor is defined as

$$K_R = \frac{R_{me}}{R_m} = K_L$$

Using these transformation factors and neglecting damping, we can write Equation (2.31) as

$$K_M m \ddot{y}(t) + K_R k y(t) = K_L p(t)$$

Dividing by K_M results in the following:

$$m \ddot{y}(t) + \frac{ky(t)}{K_{LM}} = \frac{p(t)}{K_{LM}} \tag{2.33}$$

where

$$K_{LM} = \frac{K_M}{K_L}$$

The parameter K_{LM} is defined as the load-mass factor, which is the ratio of the mass factor to the load factor. This can be a convenience because the equation can be written in terms of this one factor. Transformation factors for beams and slabs with common boundary and loading conditions are presented in tables developed more than 50 years ago (US Army Corps of Engineers 1957). Representative tables for simply supported beams and slabs are shown in Tables 2.1 and 2.2. For additional discussion and other tables of this form, the reader is referred to Biggs (1964).

Example 2.10 Determine the fundamental period of the beam shown in Figure 2.13 using transformation factors.

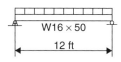

Figure 2.13

Actual System

$$\text{weight/ft} = 800 \text{ lb/ft} \quad \text{total mass} = \frac{800 \times 12}{32.2} = 298 \frac{\text{lb-sec}^2}{\text{ft}} = M$$

$$E = 30 \times 10^6 \text{ psi} \qquad I = 655.4 \text{ in}^4$$

$$\text{peak dynamic load} = 5 \text{ kips/ft} \times 12 \text{ ft} = 60 \text{ kips}$$

$$k = \frac{384EI}{5L^3} = \frac{384 \times 30 \times 10^6 \times 655.4}{5(12)^3 \times 144^2}$$

$$= 6.07 \times 10^6 \frac{\text{lb}}{\text{ft}}$$

Equivalent System

$$k_m = 0.5 \qquad m_e = 0.5 \times 298 = 149 \frac{\text{lb-sec}^2}{\text{ft}}$$

$$K_l = 0.64 \qquad p_e = 60 \times 0.64 = 38.4 \text{ kips}$$

$$k_e = kK_R = 6.07 \times 10^6 \times 0.64 = 3.88 \times 10^6$$

Period of Vibration

$$T_n = 2\pi \sqrt{\frac{149}{3.88 \times 10^6}} = 0.039 \text{ sec}$$

or

$$T_n = 2\pi \sqrt{\frac{K_{LM}M}{k}} = 2\pi \sqrt{\frac{0.78 \times 298}{6.07 \times 10^6}} = 0.039 \text{ sec}$$

Table 2.1 Transformation Factors for Beams and One-Way Slabs

Simply-supported

Loading Diagram	Strain Range	Load Factor, K_L	Mass Factor, K_M		Load-Mass Factor, K_{LM}		Maximum Resistance, R_m	Spring constant, k	Dynamic reaction, V
			Concentrated Mass*	Uniform Mass	Concentrated Mass*	Uniform Mass			
	Elastic	0.64	—	0.50	—	0.78	$\dfrac{8\mathfrak{M}_P}{L}$	$\dfrac{384EI}{5L^3}$	$0.39R + 0.11F$
	Plastic	0.50	—	0.33	—	0.66	$\dfrac{8\mathfrak{M}_P}{L}$	0	$0.38R_m + 0.12F$
	Elastic	1.0	1.0	0.49	1.0	0.49	$\dfrac{4\mathfrak{M}_P}{L}$	$\dfrac{48EI}{L^3}$	$0.78R - 0.28F$
	Plastic	1.0	1.0	0.33	1.0	0.33	$\dfrac{4\mathfrak{M}_P}{L}$	0	$0.75R_m - 0.25F$
	Elastic	0.87	0.76	0.52	0.87	0.60	$\dfrac{6\mathfrak{M}_P}{L}$	$\dfrac{56.4EI}{L^3}$	$0.525R - 0.025F$
	Plastic	1.0	1.0	0.56	1.0	0.56	$\dfrac{6\mathfrak{M}_P}{L}$	0	$0.52R_m - 0.02F$

*Equal parts of the concentrated mass are lumped at each concentrated load.

Source: "Design of Structures to Resist the Effects of Atomic Weapons," US Army Corps of Engineers Manual EM 1110-345-415, 1957.

Table 2.2 Transformation Factors for Two-Way Slabs: Simple Supports—Four Sides, Uniform Load V_A = total dynamic reaction along short edge; V_B = total dynamic reaction along long edge

Strain Range	a/b	Load Factor, K_L	Mass Factor, K_M	Load-Mass Factor, K_{LM}	Maximum Resistance	Spring Constant, k	Dynamic Reactions V_A	V_B
Elastic	1.0	0.45	0.31	0.68	$\frac{12}{a}\left(m_{Pfa}+m_{Pfb}\right)$	$\dfrac{252EI_a}{a^2}$	$0.07F+0.18R$	$0.07F+0.18R$
	0.9	0.47	0.33	0.70	$\frac{1}{a}\left(12m_{Pfa}+11m_{Pfb}\right)$	$\dfrac{230EI_a}{a^2}$	$0.06F+0.16R$	$0.08F+0.20R$
	0.8	0.49	0.35	0.71	$\frac{1}{a}\left(12m_{Pfa}+10.3m_{Pfb}\right)$	$\dfrac{212EI_a}{a^2}$	$0.06F+0.14R$	$0.08F+0.22R$
	0.7	0.51	0.37	0.73	$\frac{1}{a}\left(12m_{Pfa}+9.8m_{Pfb}\right)$	$\dfrac{201EI_a}{a^2}$	$0.05F+0.13R$	$0.08F+0.24R$
	0.6	0.53	0.39	0.74	$\frac{1}{a}\left(12m_{Pfa}+9.3m_{Pfb}\right)$	$\dfrac{197EI_a}{a^2}$	$0.04F+0.11R$	$0.09F+0.26R$
	0.5	0.55	0.41	0.75	$\frac{1}{a}\left(12m_{Pfa}+9.0m_{Pfb}\right)$	$\dfrac{201EI_a}{a^2}$	$0.04F+0.09R$	$0.09F+0.28R$
Plastic	1.0	0.33	0.17	0.51	$\frac{12}{a}\left(m_{Pfa}+m_{Pfb}\right)$	0	$0.09F+0.16R_m$	$0.09F+0.16R_m$
	0.9	0.35	0.18	0.51	$\frac{1}{a}\left(12m_{Pfa}+11m_{Pfb}\right)$	0	$0.08F+0.15R_m$	$0.09F+0.18R_m$
	0.8	0.37	0.20	0.54	$\frac{1}{a}\left(12m_{Pfa}+10.3m_{Pfb}\right)$	0	$0.07F+0.13R_m$	$0.10F+0.20R_m$
	0.7	0.38	0.22	0.58	$\frac{1}{a}\left(12m_{Pfa}+9.8m_{Pfb}\right)$	0	$0.06F+0.12R_m$	$0.10F+0.22R_m$
	0.6	0.40	0.23	0.58	$\frac{1}{a}\left(12m_{Pfa}+9.3m_{Pfb}\right)$	0	$0.05F+0.10R_m$	$0.10F+0.25R_m$
	0.5	0.42	0.25	0.59	$\frac{1}{a}\left(12m_{Pfa}+9.0m_{Pfb}\right)$	0	$0.04F+0.08R_m$	$0.11F+0.27R_m$

Source: "Design of Structures to Resist the Effects of Atomic Weapons," US Army Corps of Engineers Manual EM 1110-345-415, 1957.

2.6.4 Axial Load Effect

The generalized-coordinate approach can also be extended to include the effect of axial load. Consider the simply supported member shown in Figure 2.14 that is subjected to an axial compression load.

Because of the inclination of ds, $d\Delta = ds - dx$, and ds can be approximated as

$$ds = \sqrt{1 + \left(\frac{dv}{dx}\right)^2}\, dx \simeq 1 + \frac{1}{2}\left(\frac{dv}{dx}\right)^2$$

Substituting this approximation for ds and integrating over the length of the member, we obtain the following expression for the axial displacement:

$$\Delta = \frac{1}{2}\int_0^L v'(x)^2 dx \tag{2.34}$$

Equating the external work of the axial load to the resulting internal work gives

$$\frac{P}{2}\int_0^L v'(x)^2 dx = \frac{1}{2}\int_0^L \frac{M^2 dx}{EI} = \frac{1}{2}\int_0^L EI(x)\left[v''(x)\right]^2 dx \tag{2.35}$$

which can be written as

$$\int_0^L EI(x)\left[v''(x)\right]^2 dx - P\int_0^L \left[v'(x)\right]^2 dx = 0 \tag{2.36}$$

Introducing the generalized coordinate:

$$v'(x) = \phi'(x)y(t) \quad \text{and} \quad v''(x) = \phi''(x)y(t)$$

reduces the work equation to

$$(k_s^* - k_g^*)y(t) = 0 \tag{2.37}$$

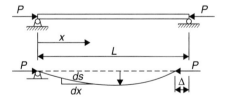

Figure 2.14 Beam with axial load

where $k_s^* = \int_0^L EI(x)[\phi''(x)]^2 dx$ = the generalized stiffness

$\quad k_g^* = P \int_0^L [\phi'(x)^2] dx$ = the generalized geometric stiffness

At the critical load, the effective stiffness equals zero and

$$P_{cr} = \frac{\int_0^L EI(x) \left[\phi''(x)\right]^2 dx}{\int_0^L [\phi'(x)]^2 dx} \qquad (2.38)$$

Example 2.11   Considering the result of Example 2.6, use the polynomial approximation for the deflected shape to determine the critical axial load if the effective stiffness is zero.

$$\phi_2(x) = 3.2 \left[\frac{x}{L} - \frac{2x^3}{L^3} + \frac{x^4}{L^4} \right]$$

$$\phi_2'(x) = 3.2 \left[-\frac{6x^2}{L^3} + \frac{x^3}{L^4} \right]$$

$$\phi_2''(x) = 38.4 \left[-\frac{x}{L^3} + \frac{x^2}{L^4} \right]$$

$$k_s^* = \int_0^L EI \left[\phi_2''(x) \right]^2 dx$$

$$= (38.4)^2 EI \int_0^L \left(\frac{x^2}{L^6} - 2\frac{x^3}{L^7} + \frac{x^4}{L^8} \right) dx = \frac{49.15EI}{L^3}$$

$$k_g^* = P \int_0^L \left[\phi'(x) \right]^2 dx = \frac{4.97P}{L}$$

and the critical load is

$$P_{cr} = \frac{9.89EI}{L^2}$$

Note that, for a simple beam, the exact critical load is the Euler load, which is given as

$$P_{cr} = \frac{\pi^2 EI}{L^2} = \frac{9.87EI}{L^2}$$

Example 2.12(M)   Solve Example 2.11 using the MATLAB Symbolic Math Toolbox.

This example involves obtaining the first- and second-order derivatives of the shape function, integrating along the length of the beam, and then dividing the results to obtain the critical load. Here is a MATLAB script to perform these operations:

```
clear all
clc
%
syms E I x L
phi=3.2*(x/L-2*x^3/L^3+x^4/L^4);    % shape function
phip = diff(phi,x,1);               % first derivative
phipp=diff(phi,x,2);                % second derivative
K=int(E*I*phipp^2,x,0,L);           % generalized stiffness
Kg=int(phip^2,x,0,L);               % geometric stiffness
Pcr = K/Kg                          % critical load
Pcr=vpa(Pcr,4)                      % show in decimal form
```

Upon execution, MATLAB will display the following results:

```
Pcr = (168*E*I)/(17*L^2)
Pcr = (9.882*E*I)/L^2
```

2.6.5 Linear Approximation

A discrete linear approximation that considers the effect of axial load is often referred to as the *string stiffness*. This representation assumes that an axial force, N, acts on a rigid bar that is connected by hinges to the flexural member. The hinges are located at points where the transverse displacement degrees of freedom of the actual member are identified and are attached to the main member by links that transmit transverse forces but no axial force components. These transverse forces depend on the value of the axial force and the slope of the segment:

$$f_{sg} = N \frac{\Delta v}{l} \qquad (2.39)$$

Substituting $\Delta v = \Delta \phi y$ and applying a virtual displacement, $\delta \Delta v = \delta \phi \Delta y$, we can rewrite the equation of motion as

$$\sum m_i \phi_i^2 \ddot{y} + c_i (\Delta \phi_i)^2 \dot{y} + \left[\sum k_i (\Delta \phi_i)^2 - \sum \frac{N}{l} (\Delta \phi_i)^2 \right] y = p^*(t) \qquad (2.40)$$

where $\sum k_i (\Delta \phi_i)^2 = $ the generalized stiffness

$\sum \frac{N}{l} (\Delta \phi_i)^2 = $ the generalized geometric stiffness

Example 2.13   Consider the simply supported beam of Example 2.6 subjected to an axial load. Use the discrete linear approach to approximate the geometric stiffness for the displaced shape, $\phi(x) = \sin\frac{\pi x}{l}$, and estimate the critical load. It is well known that the critical load for a simply supported column is

$$N_{cr} = \pi^2 \frac{EI}{L^2} = 9.87\frac{EI}{L^2}$$

Consider the beam to be divided into four equal segments. The geometric stiffness becomes

$$\sum k_{gs} = \frac{N}{l/4} \sum (\Delta\phi)_i^2$$

$$= \frac{N}{0.25l}(0.707^2 + 0.293^2 + 0.293^2 + 0.707^2) = 4.68\frac{N}{l}$$

From Example 2.6, the generalized stiffness is

$$k_s = \frac{49.15EI}{l^3}$$

Including the geometric stiffness and using the fact that, at the critical load, the effective stiffness is zero:

$$N_{cr} = 10.5\frac{EI}{l^2}$$

If the beam is divided into six segments instead of four, the critical load becomes

$$N_{cr} = 10.1\frac{EI}{l^2}$$

which compares well with the true solution:

$$N = 9.87\frac{EI}{l^2}$$

Example 2.14   Consider the four-story steel frame of Example 2.5. Use the assumed displaced shape, $\phi = \sin\frac{\pi x}{2L}$, to estimate the effect of the axial load on the lateral stiffness:

$$K^* = K_s - K_{gs} \qquad K_{gs} = \sum \frac{N_i}{h_i}(\Delta\phi_i)^2$$

Level	Story Weight, W_i (kips)	$\sum N_i$ (kips)	Story height, h_i (ft)	$\sum \dfrac{N_i}{h_i}$ (kips/in)	$\Delta\phi_i$	$\sum \dfrac{N_i}{h_i}(\Delta\phi)^2$ (kips/in)
4	294					
		294	10.5	2.33	0.071	0.012
3	368					
		662	10.5	5.25	0.203	0.216
2	368					
		1030	10.5	8.17	0.306	0.765
1	370					
		1400	12.0	9.72	0.420	1.715

$K^* = K_s - K_{gs} = 182.6 - 2.71 = 179.9$ kips/in for a reduction in stiffness of 1.5 percent.

PROBLEMS

Problem 2.1

Use d'Alembert's principle to write the equation of motion for the SDOF system shown in Figure 2.15 in terms of the indicated displacement coordinate, θ. All springs and dampers are weightless, and all displacements are small.

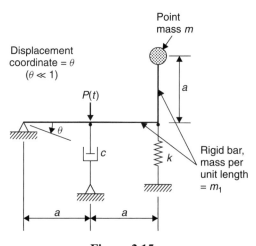

Figure 2.15

Problem 2.2

Use d'Alembert's principle to obtain the equation of motion for the system shown in Figure 2.16. Check your result using virtual work and the generalized-coordinate approach.

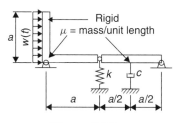

Figure 2.16

Problem 2.3

For the system shown in Figure 2.17, determine the generalized physical properties, m^*, c^*, and k^*, and the generalized loading, $p^*(t)$, in terms of the generalized coordinate, $Z(t)$. Write the equation of motion in terms of these generalized properties.

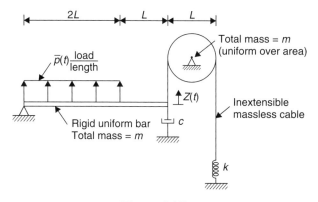

Figure 2.17

Problem 2.4

Use d'Alembert's principle to write the equation of motion for the SDOF system shown in Figure 2.18 in terms of the rotation coordinate, θ. All springs and dampers are weightless, and all displacements are small. The bar AE is rigid and has a total mass of M.

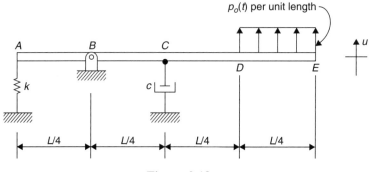

Figure 2.18

Problem 2.5

A tapered bar of constant mass density per unit volume is subjected to a support acceleration in the axial direction (see Figure 2.19). Using the generalized-coordinate approach, derive the equation of motion for the system taking into account the following two shape functions:

a. $\phi(x) = \dfrac{x}{L}$

b. $\phi(x) = \sin\dfrac{\pi x}{2L}$

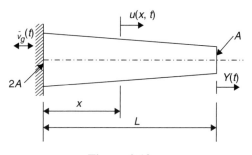

Figure 2.19

Problem 2.6

A tapered round solid cantilever beam having mass per unit volume, γ, and modulus of elasticity, E, is subjected to a dynamic force per unit length, $p(t)$, uniformly distributed along its length but varying with time (see Figure 2.20). Take the generalized coordinate as the displacement

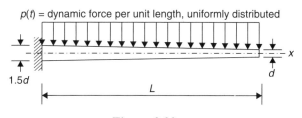

Figure 2.20

at the free end and assume that the deflected shape has the form

$$v(x,t) = \frac{1}{2}\left[3\left(\frac{x}{L}\right)^2 - \left(\frac{x}{L}\right)^3 \right] Y(t)$$

Write the equation of motion in terms of the generalized parameters.

Problem 2.7(M)

Solve Problem 2.5 using MATLAB.

Problem 2.8(M)

Solve Problem 2.6 using MATLAB.

CHAPTER 3

FREE-VIBRATION RESPONSE OF SINGLE-DEGREE-OF-FREEDOM SYSTEMS

3.1 UNDAMPED FREE VIBRATION

The term *undamped* implies there is no damping or energy dissipation present in the dynamic system, and the term *free vibration* indicates there is no applied dynamic loading. Therefore, the dynamic system consists of only a mass and a resistance, as shown in Figure 3.1. The motion of the oscillator occurs as a result of the initial conditions that occur at time zero and consist of an initial displacement and/or an initial velocity.

Summing the horizontal forces, including the inertia force, results in the equation

$$m\ddot{v} + kv = 0 \tag{3.1}$$

This equation is a linear homogeneous second-order differential equation with constant coefficients that has a general solution of the form

$$v(t) = e^{at} \tag{3.2}$$

Differentiating twice with respect to time, we obtain the following expression for the acceleration:

$$\ddot{v}(t) = a^2 e^{at} \tag{3.3}$$

Figure 3.1 Undamped single-degree-of-freedom (SDOF) linear oscillator

Now substituting Equations (3.2) and (3.3) into Equation (3.1) leads to

$$(ma^2 + k)e^{at} = 0 \qquad (3.4)$$

Because the exponential term is never zero, the expression in parentheses must be zero. Dividing by m and introducing the notation $\omega^2 = \dfrac{k}{m}$, we obtain

$$a^2 + \omega^2 = 0 \qquad (3.5)$$

which has a solution of the form $a = \pm i\omega$. Substituting this result into Equation (3.2), we get

$$v(t) = C_1 e^{i\omega t} + C_2 e^{-i\omega t} \qquad (3.6)$$

Introducing the Euler equations, $e^{\pm i\omega t} = \cos \omega t \pm i \sin \omega t$, we can write the general solution as

$$v(t) = A \sin \omega t + B \cos \omega t \qquad (3.7)$$

Differentiating Equation (3.7) leads to an equation for the velocity:

$$\dot{v}(t) = A\omega \cos \omega t - B\omega \sin \omega t \qquad (3.8)$$

Expressing the constants A and B in terms of the initial conditions at time $t = 0$ results in

$$v(0) = v_0 = B \quad \text{and} \quad \dot{v}(0) = \dot{v}_0 = A\omega \qquad (3.9)$$

and the general solution becomes

$$v(t) = \frac{\dot{v}(0)}{\omega} \sin \omega t + v(0) \cos \omega t \qquad (3.10)$$

Because the applied load is zero, the vibration of the system is initiated by these initial conditions. The solution represents a simple harmonic motion, which is shown graphically in Figure 3.2. The quantity ω is the circular frequency or angular velocity and is measured in radians per unit

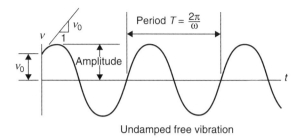

Undamped free vibration

Figure 3.2 Simple harmonic motion

of time. The period is the time it takes the displaced oscillator to make one complete cycle:

$$T = \frac{2\pi}{\omega} = \frac{1}{f} \tag{3.11}$$

The cyclic frequency, f, is the reciprocal of the period and is the number of cycles that can be completed per unit of time, normally expressed as cycles per second. It is also useful to note that 1 cycle per second is defined as 1 hertz (1 Hz).

3.1.1 Alternate Solution

The solution for the simple harmonic motion expressed in Equation (3.10) can also be written in rotating vector form as

$$\begin{aligned} v(t) &= R\cos(\omega t - \theta) \\ \dot{v}(t) &= -R\omega\sin(\omega t - \theta) \end{aligned} \tag{3.12}$$

Applying the initial conditions at $t = 0$, we obtain

$$v_0 = R\cos(-\theta) = R\cos\theta = B \tag{3.13}$$

$$\dot{v}_0 = -R\omega\sin(-\theta) = R\omega\sin\theta = A\omega$$

Squaring A and B and adding results in

$$R = \sqrt{\left(\frac{\dot{v}_0}{\omega}\right)^2 + \left(v_0\right)^2} = \text{amplitude} \qquad \theta = \tan^{-1}\frac{\dot{v}_0}{\omega v_0} = \text{phase angle}$$

It should be noted that the response is given by the real part, or horizontal projection, of the two rotating vectors given in the Argand diagram, as shown in Figure 3.3.

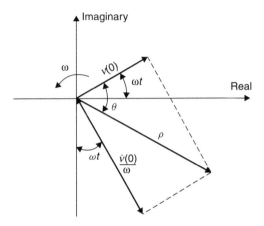

Figure 3.3 Rotating vector representation (free vibration)

In this book, "amplitude" will refer to the zero-to-peak displacement, and the "phase angle" will represent the angular distance by which the resultant motion lags behind the cosine term in the response.

Example 3.1(M) Use MATLAB to calculate and plot the first 5 sec of displacement response of an undamped SDOF linear oscillator with a natural period of $T = 1.0$ sec subjected to unit initial displacement and velocity ($v_0 = 1$, $\dot{v}_0 = 1$).
From Equation (3.11),

$$\omega = \frac{2\pi}{T} = 2\pi$$

and from Equation (3.10),

$$v(t) = \frac{\dot{v}(0)}{\omega} \sin \omega t + v(0) \cos \omega t$$

One of the most powerful features of MATLAB is that it considers every variable as a vector (or a matrix) without the need to define the dimensions of the vector beforehand. This feature will be demonstrated in this example. We will use the `linspace` function to define a set of 500 equally spaced values for t spanning from $t = 0$ to $t = 5$ sec:

```
t=linspace(0,5,500);
```

Using the MATLAB variables v0, vdot0, T, and omega to represent v_0, $\dot{v}_0$, T, and ω, respectively, we obtain the displacement response:

```
v=vdot0/omega)*sin(omega*t)+ v0*cos(omega*t);
```

Note that MATLAB variables are case-sensitive. That is why we can use t and T as different variables. We then use the plot function to plot v versus t. We could have simply used plot(t,v) to plot the response curve. However, in order to produce a better-looking graph, we specify the line width as 2 and specify the color of the line as black. The color is defined in terms of red, green, and blue (RGB), with each value of an array varying from 0 to 1. For example, [0 0 0] represents black, [1 1 1] represents white, and [1 0 0] represents red.

The complete script is as follows:

```
clear all
clc
%
% Set Initial conditions
v0=1;
vdot0=1;
%
% Define period and frequency
T=1;
omega =2*pi/T;
%
% Create an equally spaced array of 500 time values spanning
% the range of 0 to 5 seconds
t=linspace(0,5,500);
%
% Calculate the displacement response
v=vdot0/omega)*sin(omega*t)+ v0*cos(omega*t);
%
% Plot the response curve
plot(t,v,'LineWidth',2,'Color',[0 0 0]);
%
% Label the x-axis of the plot
xlabel('Time (sec)');
%
% Label the y-axis of the plot
ylabel('Displacement');
```

Upon execution, the arrays t and v are calculated, and a plot like that shown in Figure 3.4 is displayed in the MATLAB Figure 1 window.

Example 3.2 Consider the four-story steel building frame of Example 2.5 and estimate the fundamental period of the building in the transverse direction for the following four deflected shapes:

a. $\phi(x) = \sin \dfrac{\pi x}{2L}$

b. $\phi(x) = \dfrac{x}{L}$

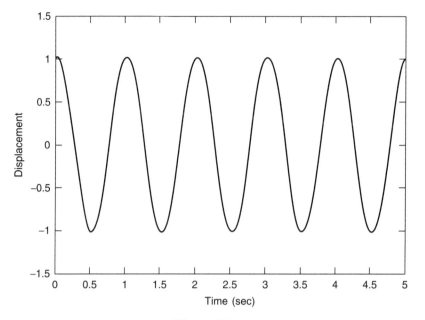

Figure 3.4

c. $\phi(x) = 1 - \cos\dfrac{\pi x}{2L}$

d. Static deflected shape

Story Stiffness (from Example 2.5)

$$K_i = \sum k_{xx} + \sum k_{yy}$$

Fourth and third stories: $K = 9(53.23) + 6(18.1) = 587.7$ kips/in
Second story: $K = 9(66.54) + 6(18.1) = 733.7$ kips/in
First story: $K = 9(44.58) + 6(15.1) = 491.6$ kips/in

a. $\phi(x) = \sin\dfrac{\pi x}{2L}$ (Example 2.4)

$$\omega = \sqrt{\frac{K^*}{M^*}} = \sqrt{\frac{182.6}{2.254}} = 9.0\ \frac{\text{rad}}{\text{sec}} \qquad T = \frac{2\pi}{\omega} = 0.698\ \text{sec}$$

b. $\phi(x) = \dfrac{x}{L}$

Level	K	M	ϕ_i	$\Delta\phi_i$	$M_i\phi_i^2$	$K_i\Delta(\phi_i)^2$
4		0.761	1.000		0.761	
	587.7			0.241		34.13
3		0.952	0.759		0.548	
	587.7			0.242		34.42
2		0.952	0.517		0.254	
	733.7			0.241		42.62
1		0.958	0.276		0.073	
	491.6			0.276		37.45
					$M^* = 1.64$	$K^* = 148.6$

$$\omega = \sqrt{\frac{148.6}{1.64}} = 9.53 \; \frac{\text{rad}}{\text{sec}} \qquad T = \frac{2\pi}{\omega} = 0.659 \text{ sec}$$

c. $\phi(x) = 1 - \cos\dfrac{\pi x}{2L}$

Level	K	M	ϕ_i	$\Delta\phi_i$	$M_i\phi_i^2$	$K_i(\Delta\phi_i)^2$
4		0.761	1.000		0.761	
	587.5			0.370		80.45
3		0.952	0.630		0.378	
	587.7			0.318		59.43
2		0.952	0.312		0.093	
	733.7			0.220		35.51
1		0.958	0.092		0.008	
	491.6			0.092		4.16
					$M^* = 1.24$	$K^* = 179.55$

$$\omega = \sqrt{\frac{179.6}{1.24}} = 12.03 \; \frac{\text{rad}}{\text{sec}} \qquad T = \frac{2\pi}{\omega} = 0.52 \text{sec}$$

d. Static deflected shape

Level	K	M	P	V	Δv	v	ϕ	$\Delta\phi$	$M_i\phi_i^2$	$K_i(\Delta\phi_i)^2$
4		0.761	100			1.733	1.00		0.761	
	587.7			100	0.170			.098		5.64
3		0.952	100			1.563	0.902		0.775	
	587.7			200	0.340			.196		22.58
2		0.952	100			1.223	0.706		0.475	
	733.7			300	0.409			.236		40.86
1		0.958	100			0.814	0.470		0.212	
	491.6			400	0.814			.470		108.59
						0.000	0.000			
									$M^* =$	$K^* =$
									2.221	177.67

$$\omega = \sqrt{\frac{177.67}{2.22}} = 8.95 \ \frac{\text{rad}}{\text{sec}} \qquad T = \frac{2\pi}{\omega} = 0.702 \text{ sec}$$

With this procedure, the assumed deflection shape functions that result in shorter periods introduce constraints into the system, which tend to shorten the period. Therefore, the best approximation to the true deflected shape is the one that has the fewest constraints and results in the longest period.

3.1.2 Rayleigh's Method

Consider the undamped oscillator shown in Figure 3.1. By proper choice of the time origin, the solution can be expressed as

$$v = v_0 \sin \omega t \qquad \dot{v} = v_0 \omega \cos \omega t \qquad \ddot{v} = -v_0 \omega^2 \sin \omega t = -\omega^2 v$$

The potential energy (PE) can be expressed as

$$\text{PE} = \frac{1}{2}kv^2 = \frac{1}{2}kv_0^2 \sin^2 \omega t$$

and the kinetic energy (KE) as

$$\text{KE} = \frac{1}{2}m\dot{v}^2 = \frac{1}{2}mv_0^2\omega^2 \cos^2 \omega t$$

By the conservation of energy (Newton),

$$\text{total energy (TE)} = \text{KE}_{max} = \text{PE}_{max} = \text{constant} \qquad (3.14)$$

Substituting for KE_{max} and PE_{max} and simplifying, we get the following expression for the circular frequency:

$$\omega^2 = \frac{k}{m} \qquad \text{and} \qquad T = \frac{2\pi}{\omega} \qquad (3.15)$$

This energy balance of Rayleigh can also be recast in terms of a generalized-coordinate system:

$$v(x,t) = \phi(x)Y(t) = \phi(x)Y_0 \sin \omega t \qquad v(x)_{max} = \phi(x)Y_0$$

$$\dot{v}(x,t) = \phi(x)Y_0 \omega \cos \omega t \qquad \dot{v}(x)_{max} = \phi(x)Y_0 \omega$$

Substituting into the term for internal potential energy results in

$$\text{PE}_{max} = \frac{1}{2}\sum k_i v_i^2 = \frac{1}{2}k^* Y_0^2$$

which can also be expressed in terms of external work as

$$\text{PE}_{max} = \frac{1}{2}\sum P_i v_i = \frac{1}{2}P^* Y_0 \qquad (3.16)$$

$$\text{KE}_{max} = \frac{1}{2}\sum m_i \dot{v}_i^2 = \frac{1}{2}m^* \omega^2 Y_0^2 \qquad (3.17)$$

Equating KE_{max} to PE_{max} and simplifying, we get an expression for the circular frequency in terms of the generalized parameters:

$$\omega^2 = \frac{P^*}{m^* Y_0} \qquad (3.18)$$

The period of vibration can be developed in a similar manner starting with

$$T = \frac{2\pi}{\omega} = 2\pi\sqrt{\frac{m^* Y_0}{P^*}} = 2\pi\sqrt{\frac{Y_0 \sum m_i \phi_i^2}{\sum P_i \phi_i}} \qquad (3.19)$$

Now multiply the top and bottom by Y_0 and substitute $\delta_i = \phi_i Y_0$ and $m_i = \dfrac{w_i}{g}$ to obtain

$$T = 2\pi\sqrt{\frac{\sum w_i \delta_i^2}{g \sum P_i \delta_i}} \qquad (3.20)$$

where δ_i are the elastic deflections calculated using the applied lateral forces, P_i, which are distributed over the structure in a rational manner. This is the equation that is given in building codes for estimating the fundamental period of a building.

3.1.3 Selection of Deflected Shape

1. The lowest frequency (longest period) is always the best approximation. If the selected shape is not the true shape, constraints are being imposed on the dynamic system. These constraints will raise the frequency and shorten the period.
2. The best approximation to the deformed shape is obtained when a rational pattern of external loads is applied and the resulting deflection is calculated.

Example 3.3 Discrete Load

Internal strain energy: $\mathrm{PE}_{\max} = \dfrac{1}{2}k^*Y_0^2$

Potential energy: $\mathrm{PE}_{\max} = \dfrac{1}{2}P^*Y_0$

Kinetic energy: $\mathrm{KE}_{\max} = \dfrac{1}{2}m^*\omega^2 Y_0 = \dfrac{1}{2}P^*Y_0^2$

The circular frequency is then given as

$$\omega^2 = \frac{P^*}{m^*Y_0}$$

Example 3.4 Distributed Load

Potential energy: $\mathrm{PE}_{\max} = \dfrac{1}{2}Y_0 \displaystyle\int_0^L q(x)\phi(x)dx$

Kinetic energy: $\mathrm{KE}_{\max} = \dfrac{1}{2}m^*\omega^2 Y_0^2$

The circular frequency is then given as

$$\omega^2 = \frac{\int_0^L q(x)\phi(x)dx}{m^*Y_0} = \frac{P^*}{m^*Y_0}$$

3.2   DAMPED FREE VIBRATION

If some form of energy dissipation such as viscous damping is introduced into the dynamic system, as shown in Figure 3.5, the equation of motion has the form

$$m\ddot{v} + c\dot{v} + kv = 0 \tag{3.21}$$

which can be written as

$$\ddot{v} + \frac{c}{m}\dot{v} + \omega^2 v = 0 \tag{3.22}$$

The solution to this differential equation has the form $v = e^{at}$. Substituting this value of v and its derivatives into the equation of motion results in the auxiliary equation

$$a^2 + \frac{c}{m}a + \omega^2 = 0 \tag{3.23}$$

which has the roots

$$a = -\frac{c}{2m} \pm \sqrt{\left(\frac{c}{2m}\right)^2 - \omega^2} \tag{3.24}$$

Three possible solutions are represented by this equation depending on whether the quantity under the square root sign is positive, negative, or zero. The general solution has the form

$$v(t) = Ae^{at} + Be^{bt} \tag{3.25}$$

1. *Critical damping:* The limiting condition occurs when the radical is zero. Critical damping is defined as the minimum viscous damping that will allow a displaced system to return to its initial position without oscillation. This condition implies that

$$\frac{c}{2m} = \omega \quad \text{or} \quad c_{\text{cr}} = 2m\omega$$

Figure 3.5   Damped SDOF oscillator

and

$$a = -\frac{c}{2m} = -\omega \text{ (repeated root)}$$

which results in the general solution $v(t) = (A + Bt)e^{-\omega t}$. The actual damping in a structural system is usually estimated in terms of a new parameter defined as the percentage or fraction of critical damping, which is defined as

$$\xi = \frac{c}{c_{cr}}$$

2. *Radical negative:* In this case,

$$\left(\frac{c}{2m}\right)^2 - \omega^2 < 0$$

implies

$$c < 2m\omega = c_{cr} \quad \text{and} \quad c = 2\xi m\omega = \xi c_{cr}$$

$$\frac{c}{2m} = \xi\omega \quad \text{and} \quad a = -\xi\omega \pm \sqrt{(\xi\omega)^2 - \omega^2}$$

$$= -\xi\omega \pm i\omega\sqrt{1 - \xi^2} = -\xi\omega \pm i\omega_D$$

where $\omega_D = \omega\sqrt{1 - \xi^2}$ = the damped circular frequency

For small values of damping normally found in structural systems, $\omega_D \cong \omega$; however, in structural systems where supplemental damping in employed, this may not be the case.

Example 3.5 Effect of Damping on Natural Frequency If $\xi = 5\%$ $= 0.05$, $\omega_D = 0.9987\omega \cong \omega$. However, if $\xi = 25\% = 0.25$, $\omega_D = 0.9682\omega$. Substituting into the general solution:

$$v(t) = Ae^{-\xi\omega t + i\omega_D t} + Be^{-\xi\omega t - i\omega_D t} = e^{-\xi\omega t}\left(Ae^{i\omega_D t} + Be^{-i\omega_D t}\right) \tag{3.26}$$

Using Euler's equations, $e^{i\theta} = \cos\theta + i\sin\theta$ and $e^{-i\theta} = \cos\theta - i\sin\theta$, we obtain the following solution:

$$v(t) = e^{-\xi\omega t}\left(A\cos\omega_D t + Ai\sin\omega_D t + B\cos\omega_D t - Bi\sin\omega_D t\right) \tag{3.27}$$

This expression can be written in a more convenient form as

$$v(t) = e^{-\xi \omega t} (C \sin \omega_D t + D \cos \omega_D t) \qquad (3.28)$$

The term in parentheses represents harmonic motion at the constant circular frequency, ω_D, and the exponential term represents a decay term that dampens out the response with time. The constants are determined from the initial conditions:

$$D = v_0 \quad \text{and} \quad C = \frac{\dot{v}_0 + v_0 \xi \omega}{\omega_D}$$

Applying the initial conditions to the main equation results in the following:

$$v(t) = e^{-\xi \omega t} \left[\frac{\dot{v}_0 + v_0 \xi \omega}{\omega_D} \sin \omega_D t + v_0 \cos \omega_D t \right] \qquad (3.29)$$

3.2.1 Rotating Vector Form

$$v(t) = Re^{-\xi \omega t} \cos (\omega_D t - \phi) \qquad (3.30)$$

$$R = \left[\left(\frac{\dot{v}_0 + v_0 \xi \omega}{\omega_D} \right)^2 + v_0^2 \right]^{1/2} \qquad (3.31)$$

$$\phi = \tan^{-1} \left(\frac{\dot{v}_0 + v_0 \xi \omega}{\omega_D v_0} \right) \qquad (3.32)$$

Example 3.6(M) Use MATLAB to calculate and plot the first 5 sec of displacement response of a damped SDOF linear oscillator similar to that of Example 3.1(M), assuming (a) $\xi = 0.05$ and (b) $\xi = 0.50$. In each case, show that $Re^{-\xi \omega t}$ envelops the response curve.

All we need to do is to modify the script developed for Example 3.1(M) to introduce damping and add a second curve to the plot for $Re^{-\xi \omega t}$. We will use the rotating vector form and have MATLAB prompt us for the value of ξ. Notice that we are plotting the response curve and positive and negative envelope curves via a single command and then changing the line style of the response curve. The script is as follows:

```
clear all
clc
```

```
%
zeta = input('Fraction of Critical Damping = ');
v0=1;
vdot0=1;
T=1;
omega =2*pi/T;
omegad=omega*sqrt(1-zeta^2)
%
t=linspace(0,5,1000);
%
arg1= exp(-zeta*omega*t)
%
R=sqrt(((vdot0+v0*zeta*omega)/omegad)^2 + v0^2);
phi=atan((vdot0+v0*zeta*omega)/(omegad*v0));
% IMPORTANT NOTE:
%   we must use dot product of vectors in the following two statements
%   (i.e. ".*" instead of "*". Otherwise we get vector size incompatibility
%   error.
Env=R.*arg1;
v=Env.*cos(omegad*t-phi)
%
% Create plots
plot1=plot(t,v,t,Env,t,-Env,'LineWidth',1,'Color',[1 0 0]',
'LineStyle','--');
set(plot1(1),'LineWidth',2,'Color',[0 0 0],'LineStyle','-');
%
% Create xlabel
xlabel('Time (sec)');
%
% Create ylabel
ylabel('Displacement');
```

Upon execution, we are prompted for the value of ξ:

```
Fraction of Critical Damping =
```

Executing the script twice for ξ:

a. $\xi = 0.05$ results in calculation of the response and display of the graph shown in Figure 3.6.

b. $\xi = 0.50$ results in calculation of the response and display of the graph shown in Figure 3.7.

Notice how the increase in damping has resulted in a rapid reduction of the amplitude of displacement and the number of cycles of response with measurable amplitude.

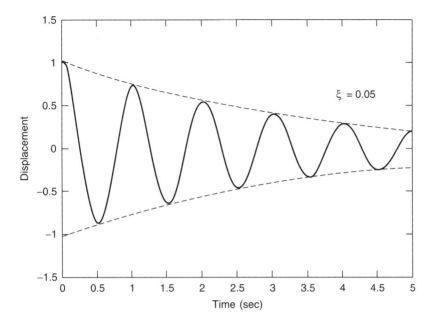

Figure 3.6

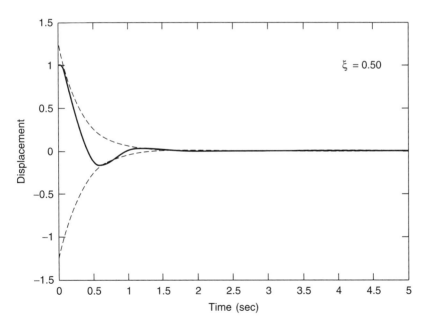

Figure 3.7

3.2.2 Logarithmic Decrement

Free vibration of a dynamic system provides one means of determining the fraction (%) of critical damping, ξ. A plot of the response of a system with less than critical damping to an initial displacement with zero initial velocity is shown in Figure 3.8.

The general solution has the form

$$v(t) = Re^{-\xi \omega t} \cos(\omega_D t - \phi) \tag{3.33}$$

If $\cos(\omega_D t - \phi) = 1$, $v_1(t) = Re^{-\xi \omega t}$ and because the time interval between successive peaks is a constant, $T_D = \dfrac{2\pi}{\omega_D}$, the displacement at the second peak can be written as

$$v_2(t) = Re^{-\xi \omega (t + 2\pi/\omega_D)} \tag{3.34}$$

The ratio between successive peaks becomes

$$\frac{v_1}{v_2} = e^{\xi \omega (2\pi/\omega_D)}$$

and the natural logarithm of this ratio is called the *logarithmic decrement*, δ:

$$\delta = \ln \frac{v_1}{v_2} = 2\pi \xi \frac{\omega}{\omega_D} \tag{3.35}$$

For lightly damped systems with $\omega \cong \omega_D$,

$$\delta = 2\pi \xi \quad \text{or} \quad \xi = \frac{\delta}{2\pi} \tag{3.36}$$

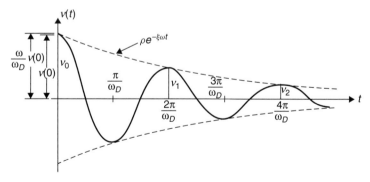

Figure 3.8 Damped free vibration

If $\omega \neq \omega_D$,

$$\delta = \xi \omega T_D = \frac{2\pi \xi}{\left(1 - \xi^2\right)^{1/2}}$$

$$v(t) = e^{-\xi \omega t} (A \sinh \hat{\omega}t + B \cosh \hat{\omega}t) \tag{3.37}$$

$$\xi = \frac{\delta/2\pi}{\left[1 + (\delta/2\pi)^2\right]^{1/2}} \tag{3.38}$$

3.2.3 Radical Positive

If the damping ratio is greater than the critical damping ratio, the system is said to be overdamped:

$$\xi > \xi_{cr} = 2m\omega \Rightarrow \sqrt{\frac{c}{2m} - \omega^2} > 0$$

$$a = -\xi \omega \pm \omega \sqrt{\xi^2 - 1} = -\xi \omega \pm \hat{\omega}$$

where $\hat{\omega} = \omega \sqrt{\xi^2 - 1}$

$$v(t) = e^{-\xi \omega t} (A \sinh \hat{\omega}t + B \cosh \hat{\omega}t) \tag{3.39}$$

The effect of various damping ratios on the response of the system is shown in Figure 3.9.

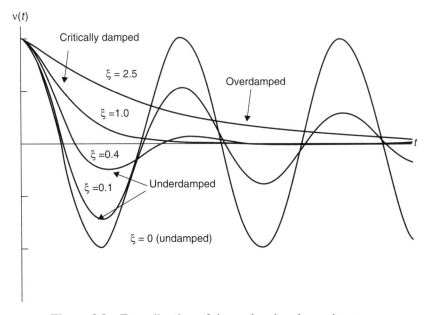

Figure 3.9   Free vibration of damped and undamped systems

Example 3.7 The SDOF oscillator shown in Figure 3.5 is set in motion by an initial displacement, $v(0) = 1$. The ratio of the initial displacement to the succeeding displacement is 1.18. The oscillator has the following properties: $m = W/g = 10$ lb/386.4 in/sec and $k = 20$ lb/in.
Determine the following:

a. The natural circular frequency, ω; the natural cyclic frequency, f; and the fundamental period, T:

$$\omega = \sqrt{\frac{k}{m}} = \sqrt{\frac{20 \times 386.4}{10}} = 27.8 \ \frac{\text{rad}}{\text{sec}}$$

$$f = \frac{\omega}{2\pi} = \frac{27.8}{2\pi} = 4.42 \ \text{Hz}$$

$$T = \frac{1}{f} = \frac{1}{4.42} = 0.23 \ \text{sec}$$

b. The logarithmic decrement:

$$\delta = \ln \frac{v_1}{v_2} = \ln(1.18) = 0.163$$

c. The damping ratio:

$$\xi = \frac{\delta}{2\pi} = \frac{0.163}{2\pi} = 0.026 = 2.6\%$$

d. The damping coefficient:

$$c = \xi c_{\text{cr}} = 2\xi\omega m = 2 \times 0.026 \times 27.8 \times \frac{10}{386.4} = 0.037 \ \text{lb-sec/in}$$

e. The damped natural frequency:

$$\omega_D = \omega\sqrt{1 - \xi^2} = 27.8\sqrt{1 - (0.026)^2}$$

$$= 27.8(0.9993) = 27.78 \cong \omega$$

Example 3.8(M) Use MATLAB to directly solve the differential equation of motion [Equation (3.21)] for the SDOF oscillator of Example 3.7 for damping ratios $\xi = 0.0, 0.10, 0.40, 1.0$, and 2.5 for the time span of 0 to 1 sec.

Given $c = 2\xi m\omega$, Equation (3.21) can be rewritten as

$$\ddot{v} + 2\xi\omega\dot{v} + \omega^2 v = 0$$

where $\omega = 27.8$ rad/sec, $v_0 = 1$, and $\dot{v}_0 = 0$.

MATLAB has several built-in functions for solving ordinary differential equations. All these functions return the numerical solution to a system of first-order differential equations. The most frequently used function of this type for initial value problems is ode45, which we will use in this example. At a first glance, we appear to be out of luck because ode45 solves first-order differential equations while our equation is a second-order one. However, we can convert our second-order differential equation to a set of two first-order differential equations by making the following substitutions:

$$v_1 = v$$

$$v_2 = \dot{v}$$

Converting our equation to the following two first-order equations, we have

$$\frac{dv_1}{dt} = v_2$$

$$\frac{dv_2}{dt} = -2\xi\omega v_2 - \omega^2 v_1$$

Now we can use ode45 to solve the previous system of first-order equations. We can use one of the following syntaxes for ode45:

```
[T,Y] = ode45
(odefun,tspan,y0)[T,Y] = ode45
(odefun,tspan,y0,options)[T,Y,TE,YE,IE] = ode45
 (odefun,tspan,y0,options)
sol = ode45(odefun,[t0 tf],y0...)
```

where

odefun is a function handle that evaluates the right side of the differential equations

tspan is a vector specifying the interval of integration, [t0,tf]. The user imposes the initial conditions at tspan(1) and integrates from tspan(1) to tspan(end)

y0 is a vector of initial conditions

T is the column vector of time points

Y is the solution array. Each row in Y corresponds to the solution at a time returned in the corresponding row of T

We need to define our external function for odefun. We define our function, DLSDOF, as follows:

```
function v = DLSDOF (t, v, zeta)
v= [v(2); -2*27.8*zeta*v(2)-27.8*27.8*v(1)];
```

Our ode45 call would look like the following:

```
[t, v] = ode45(@DLSDOF, tspan, [1 0]', zeta(n))
```

Note that [1 0]' in the previous statement is the transpose of the row vector [1 0] representing our initial conditions. The complete script is as follows:

```
close all
clear
clc
%
zeta = [0.0, 0.10, 0.40, 1.0, 2.5];
tspan = linspace(0, 1, 100);
for n = 1:5
    [t, v] = ode45(@DLSDOF, tspan, [1 0]', [], zeta(n))
    plot(t, v(:,1));
    hold on
end
xlabel('Time (seconds)')
ylabel('Displacement (in)')
axis([0,1.0,-1.5,1.5]);
plot([0,1.0],[0,0],'k-')
function v = DLSDOF (t, v, zeta)
v= [v(2); -2*27.8*zeta*v(2)-27.8*27.8*v(1)];
```

Upon execution, the responses for various damping ratios are calculated and reported on-screen, and a graph of the results is displayed, as shown in Figure 3.10.

You can use the Plot Tool to add legends and change colors and line types to improve the appearance and the information that the graph conveys (see Figures 3.11 and 3.12).

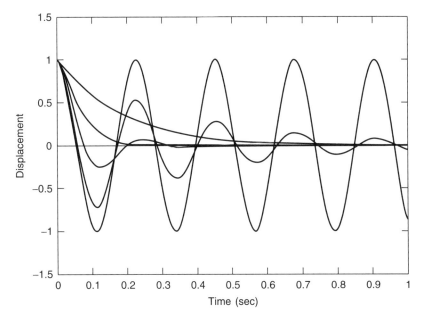

Figure 3.10

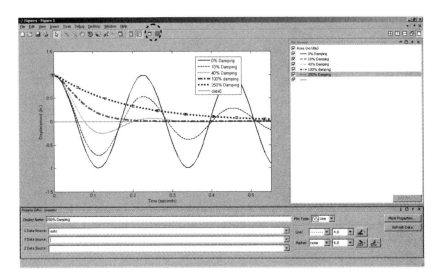

Figure 3.11

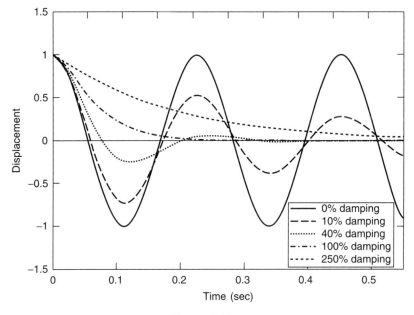

Figure 3.12

As we will see in the following chapters, we can use modified versions of this basic script to obtain the responses of linear SDOF systems to all kinds of excitations.

PROBLEMS

Problem 3.1

Estimate the fundamental period of vibration for translation of the six-story steel office building shown in Figure 3.13. Assume the floor diaphragm is rigid in its own plane and that the beams are rigid relative to the columns (compare I_g to I_c). Note that the building framing is approximately the same in both directions so the periods should also be approximately the same. The foundation for each column of the perimeter frame consists of two 30 in diameter piles that are 32 ft in length. The weight of the structure should be estimated based on the unit weights given in Figure 3.14.

Notes: There is a 5 ft overhang of the floor deck at the second floor level and a 6.5 ft overhang at the roof level. The equipment penthouse

Roof:

Roof Deck . . . . 46.0 psf

      20 gauge metal deck 3 in deep with $3\frac{1}{4}$ in lightweight
      concrete on top

Roofing . . . . 6.0 psf

Hung ceiling . . . . 8.0 psf

Mechanical equipment penthouse . . . . 43.0 psf

Steel . . . . 8.0 psf

Typical Floor:

Floor deck . . . . 46.0 psf

      (same as roof deck)

Hung ceiling . . . . 8.0 psf

Floor finish . . . . 1.0 psf

Partitions . . . . 15.0 psf

Steel . . . . 10.8 psf

      (Beams and joists . . . 7.1 psf, columns . . . 3.7 psf)

Perimeter Wall:

Glass with mullions . . . . 10.0 psf

Figure 3.13

is approximately 46 ft × 62 ft and is located in the center of the building plan.

Problem 3.2

The single-bay moment-resistant steel frame shown in Figure 3.15 has the given properties. The moment of inertia shown is for each column at a story level, and the beams are assumed to be rigid relative to the columns. Use a generalized coordinate of unity at the roof level and the static deflected shape to estimate the period of vibration of the frame.

Problem 3.3(M)

Solve Problem 3.2 using MATLAB.

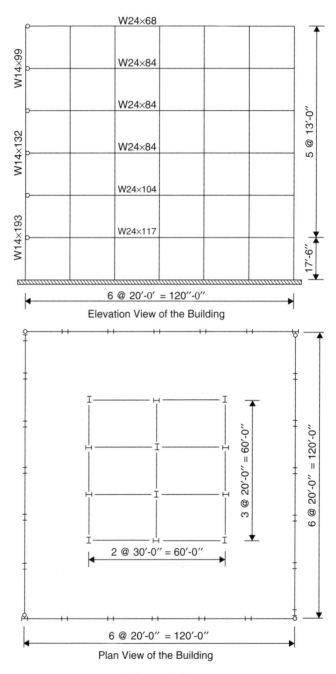

Elevation View of the Building

Plan View of the Building

Figure 3.14

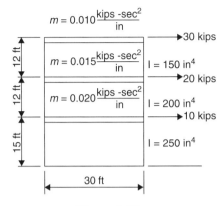

Figure 3.15

Problem 3.4

Estimate the fundamental period of vibration of the office building shown in Figure 3.16. The roof weight is 1400 kips, and the two floors each weigh 3200 kips. Use the following deflected shapes:

  a. Static deflected shape

  b. $\phi(x) = \sin \dfrac{\pi x}{2L}$ and $\phi(x) = \dfrac{x}{L}$

Assume pinned connections for the girders and pipe bracing.

Problem 3.5(M)

Solve Problem 3.4 using MATLAB.

Problem 3.6

The typical exterior moment frame of a three-story steel building is shown in Figure 3.17. The roof mass is 35.5 kip-sec^2/ft, and the masses for the two floors are both 32.8 kip-sec^2/ft. Use the static deflected shape to estimate the fundamental period of this building.

Problem 3.7(M)

Solve Problem 3.6 using MATLAB.

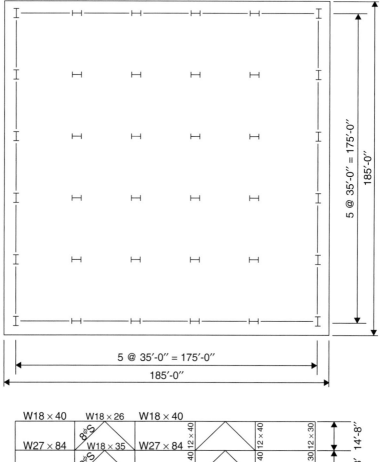

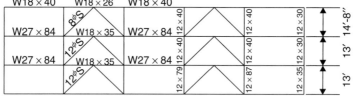

Figure 3.16

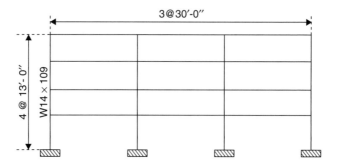

Figure 3.17

CHAPTER 4

RESPONSE TO HARMONIC LOADING

4.1 UNDAMPED DYNAMIC SYSTEM

An undamped dynamic system subjected to a harmonically varying load has an equation of motion of the form

$$m\ddot{v} + kv = p_0 \sin pt \qquad (4.1)$$

where p_0 = the amplitude of the driving force

p = the driving frequency of the harmonic load, which is generally not the same as the natural frequency, ω

The general solution of this differential equation consists of two parts as given by the following:

$$v(t) = v_h(t) + v_p(t) \qquad (4.2)$$

where $v_h(t) = A \sin \omega t + B \cos \omega t$ = the homogeneous solution that gives the free-vibration response

$v_p(t)$ = the particular solution that gives the behavior generated by the form of the dynamic loading

The particular solution has the form $v_p(t) = D \sin pt$, where the constant D is determined from the requirement that the particular solution must satisfy the equation of motion. Taking the form of the particular solution and substituting into the equation of motion, we have

$$-Dp^2 m \sin pt + Dk \sin pt = p_0 \sin pt$$

which results in the following expression for the constant D:

$$D = \frac{p_0}{k(1 - \beta^2)} \tag{4.3}$$

where $\beta = \frac{p}{\omega}$ is the ratio of the applied load frequency to the natural frequency of the system.

Substituting the value for D, we get the following general solution for velocity and displacement:

$$\dot{v}(t) = A\omega \cos \omega t - B\omega \sin \omega t + \frac{p_0}{k} \left(\frac{1}{1 - \beta^2} \right) p \cos pt$$

$$v(t) = A \sin \omega t + B \cos \omega t + \frac{p_0}{k} \left(\frac{1}{1 - \beta^2} \right) \sin pt \tag{4.4}$$

As before, the constants A and B depend on the initial conditions. In the case of a system starting from rest, the initial conditions are $v(0) = \dot{v}(0) = 0$. Applying these conditions, the constants take the form

$$A = -\frac{p_0}{k} \beta \left(\frac{1}{1 - \beta^2} \right) \quad \text{and} \quad B = 0$$

Applying these constants to the general form of the solution results in

$$v(t) = \frac{p_0}{k} \left(\frac{1}{1 - \beta^2} \right) (\sin pt - \beta \sin \omega t) \tag{4.5}$$

where the individual terms can be identified as follows:

$\frac{p_0}{k} = $ static displacement $\frac{1}{1 - \beta^2} = $ amplification factor

$\sin pt = $ steady-state response $\beta \sin \omega t = $ transient response

From Equation (4.5), it can be seen that, for lightly damped systems, the peak steady-state response occurs at a frequency ratio near unity when the exciting frequency of the applied load equals the natural frequency of the system. This condition is called *resonance*. The ratio of the dynamic displacement to the static displacement is often referred to as the *response ratio* or the *dynamic load factor* (DLF) and is defined as follows:

$$R(t) = \frac{v(t)}{v_{\text{static}}} = \frac{1}{1 - \beta^2}(\sin pt - \beta \sin \omega t)$$

$$= \text{dynamic load factor (response ratio)} \qquad (4.6)$$

This equation implies that the response of the undamped system goes to infinity at resonance; however, a closer examination by Clough and Penzien (1975) in the region of β equal to unity shows that it only tends toward infinity and that several cycles are required for the response to build up. Hence, duration is an important consideration for this type of motion. Starting with Equation (4.5) and letting $\omega - p = 2\Delta\omega$, $\beta \cong 1.0$, $\sin \alpha - \sin \beta = 2 \cos \frac{\alpha+\beta}{2} \sin \frac{\alpha-\beta}{2}$, and $\sin \Delta\omega t = \Delta\omega t$ as $\Delta\omega t \to 0$, the following result is obtained, after making the indicated substitutions and simplifying,

$$v(t) = \frac{v_{\text{static}}\omega t}{2} \cos \omega t \qquad (4.7)$$

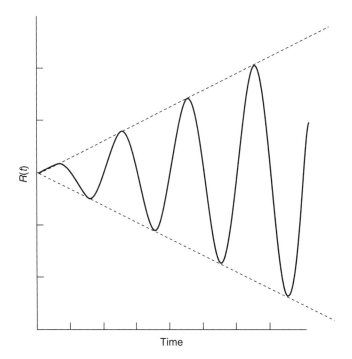

Time

Figure 4.1 Resonance response for an undamped oscillator

A plot of the response ratio $R(t)$ versus time is shown in Figure 4.1 for an undamped oscillator.

Example 4.1(M)   Use MATLAB to calculate and plot the response ratio $R(t)$ versus time for the first 5 sec of the response of an undamped system with a natural period of 1.0 sec subjected to a harmonic forcing function with a frequency of 1.0 Hz. What is the maximum value of $R(t)$ in this plot?

$$\omega = \frac{2\pi}{T} = 2\pi \quad \text{and} \quad p = 2\pi f = 2\pi$$

Therefore, $\omega = p$, and the system is in resonance:

$$v(t) = \frac{v_{\text{static}}\omega t}{2} \cos \omega t$$

Therefore,

$$R(t) = \frac{v(t)}{v_{\text{static}}} = \frac{\omega t}{2} \cos \omega t$$

All we have to do is to write a simple script to calculate and plot $R(t)$. We define time as an array starting at 0 and ending at 5 sec, with an increment of 0.01 sec, and ω as a scalar:

```
t=0:0.01:5.0;
omega = 2*pi
```

The vector containing values of $R(t)$ is obtained from

```
R=0.5*omega.*t.*cos(omega*t);
```

The maximum values of $R(t)$ in the plot are obtained from

```
Rmax=max(R)
```

The entire MATLAB script is as follows:

```
clear all
clc
%
omega = 2*pi
%
t=0:0.01:5.0;
%
% Note that in the following statement we are using dot products of
```

```
% vectors ".*" instead of "*" to multiply scalars by vectors and
% vectors by vectors. Failure to do so will result in error
messages.
%
R=0.5*omega.*t.*cos(omega*t);

Rmax=max(R)
plot(t,R)
grid on
xlabel('t (sec)')
ylabel('Response Ratio, R(t)')
xlim([0 5])
```

Upon execution, we obtain the graph for $R(t)$ shown in Figure 4.2 and its maximum plotted value:

```
Rmax = 15.7080
```

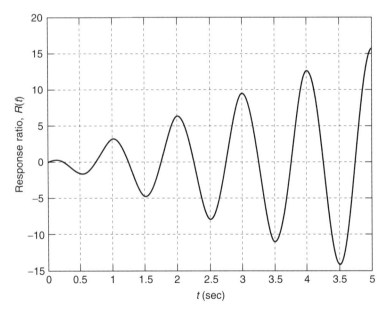

Figure 4.2

Consider a harmonic load in which the forcing frequency is approximately equal to the natural frequency of the dynamic system, $p \cong \omega$. Using the definition of the response ratio given in Equation (4.6) for an undamped (lightly damped) dynamic system, we can write the dynamic displacement as

$$v(t) = v_{\text{static}} \left(\frac{1}{1 - \beta^2} \right) (\sin pt - \beta \sin \omega t) \qquad (4.8)$$

Now let $\omega - p = 2\Delta\omega$ and $\beta \cong 1$. Using the identity

$$\sin \alpha - \sin \beta = 2 \cos \frac{\alpha + \beta}{2} \sin \frac{\alpha - \beta}{2}$$

we can express the dynamic displacement as

$$v(t) = v_{\text{static}} \frac{\omega^2}{\omega^2 - p^2} \sin \Delta\omega t \cdot 2 \cos \left(\frac{p + \omega}{2} \right) t$$

Noting that $\dfrac{p + \omega}{2} \cong \omega \gg \Delta\omega$ and that $\omega + p \cong 2\omega$ results in the following expression for the displacement response:

$$v(t) = \frac{v_{\text{static}}\omega}{2\Delta\omega} \sin \Delta\omega t \cos \omega t \qquad (4.9)$$

The total response is the sum of two terms, one that represents the contribution of the dynamic loading and is referred to as *steady state*, and the other that represents the contribution of the initial conditions and is referred to as *transient*. The tendency of these two components to get in phase and then out again causes a response like that shown in Figure 4.3. This response is called *beating* and is caused by two harmonic oscillations with slightly different frequencies.

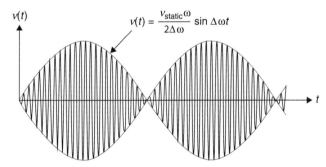

Figure 4.3 Response of two harmonic oscillations with slightly different frequencies

Example 4.2(M) An undamped system with a natural frequency of 2.0 Hz is subjected to a harmonic forcing function with a frequency of 2.08 Hz. The static deflection under the load is 1.0 in. Use MATLAB to calculate and plot and compare the first 14 sec of the displacement response of the system obtained by Equations (4.8) and (4.9).

Here is the MATLAB script for solving this problem. Note that we are plotting the results of Equations (4.8) and (4.9) with solid lines and dashed lines, respectively.

```
clear all
clc
%
f=2.0;
ff=2.08;
omega = 2*pi*f;
p=2*pi*ff;
vstatic=1;
beta= p/omega
%
% Set the time vector
%
t=0:0.01:14;
%
% Using Equation 4.8
%
v48=(vstatic/(1-beta^2))*(sin(p*t)-beta*sin(omega*t));
%
% Using Equation 4.9
%
DeltaOmega= (omega-p)/2;
wr = omega/DeltaOmega;
% Use dot products of vectors. Failure to do so will result
% in error messages.
v49=(0.5*vstatic*wr).*sin(DeltaOmega*t).*cos(omega*t);
plot(t,v48,'-k',t,v49,'--r')

grid on
xlabel('t (sec)')
ylabel('Displacement Response')
```

Upon execution of the script, a graph like that shown in Figure 4.4 will be displayed. Close examination of the graph reveals that, although the results obtained by the two equations are very close, there is a noticeable phase lag between the two curves.

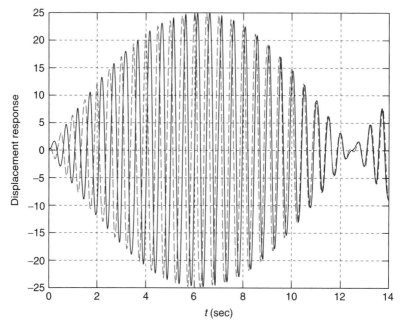

Figure 4.4

4.2 DAMPED DYNAMIC SYSTEM

The more general solution to the harmonic excitation includes the addition of a term representing viscous damping, and the equation of motion becomes

$$m\ddot{v} + c\dot{v} + kv = p_0 \sin pt \qquad (4.10)$$

As in the undamped case, the general solution of this differential equation consists of the sum of a complementary solution representing the free-vibration response and a particular solution representing the dynamic loading. The damped free-vibration response is given by

$$v_h(t) = e^{-\xi \omega t} (A \sin \omega_D t + B \cos \omega_D t) \qquad (4.11)$$

and the particular solution for the harmonic loading has the form

$$v_p(t) = C_1 \sin pt + C_2 \cos pt \qquad (4.12)$$

The particular solution must satisfy the equation of motion. Therefore, taking the first and second derivatives of Equation (4.12) and separating the multiples of $\sin pt$ and $\cos pt$, we obtain the following set of equations:

$$\left[-mC_1p^2 - cC_2p + kC_1\right]\sin pt = p_0 \sin pt \qquad (4.13)$$

$$\left[-mC_2p^2 + cC_1p + kC_2\right]\cos pt = 0$$

Solving these equations simultaneously, we get the following values for C_1 and C_2:

$$C_1 = \frac{p_0}{k}\left[\frac{1-\beta^2}{(1-\beta^2)^2 + (2\xi\beta)^2}\right] \qquad (4.14)$$

$$C_2 = \frac{p_0}{k}\left[\frac{-2\xi\beta}{(1-\beta^2)^2 + (2\xi\beta)^2}\right]$$

Introducing these coefficients into the particular solution and combining it with the complementary solution, we obtain the general solution:

$$v(t) = e^{-\xi\omega t}\left(A\sin\omega_D t + B\cos\omega_D t\right)$$

$$+ \frac{p_0}{k}\frac{1}{(1-\beta^2)^2 + (2\xi\beta)^2}\left[(1-\beta^2)\sin pt - 2\xi\beta\cos pt\right] \qquad (4.15)$$

The first term represents the transient response to the applied loading. The coefficients A and B can be evaluated for any initial conditions; however, for a system with any damping, this term dampens out after a limited number of cycles and is of little interest. The second term is the steady-state response at the frequency of the applied loading but out of phase with it.

An alternative representation of the motion may be obtained in a plot of the displacement versus the velocity in the so-called phase plane. The displacement is plotted as the ordinate, and the velocity divided by the frequency of the driving force is plotted as the abscissa, as shown in Figure 3.3. The two terms in the particular solution to Equation (4.12) appear as rotating vectors in Figure 3.3. The sine term is a vector of length C_1 at an angle pt above the horizontal axis, and the cosine term is a vector of length C_2 perpendicular to the C_1 vector. This indicates that the C_1 vector lags behind the C_2 vector by a constant angle of $\pi/2$ rad. The two vectors of unchanging length are always perpendicular and rotate counterclockwise in the phase plane at a constant angular velocity of p rad/sec. From the geometry given in Figure 3.3, it can be seen that the displacement can be represented as

$$v = \left(C_1^2 + C_2^2\right)^{1/2} \sin(pt - \phi) = R \sin(pt - \phi) \tag{4.16}$$

where $R = \dfrac{p_0}{k} R_d$

$$R_d = \frac{1}{\sqrt{[1 - \beta^2]^2 + [2\xi\beta]^2}} = \text{the displacement magnification factor}$$

$v = \dfrac{p_0}{k} R_d \sin(pt - \phi)$, where ϕ is the phase angle by

which the response lags behind the applied load. (4.17)

It can be seen from Equation (4.15) that, at resonance (i.e., $\beta = 1$), the DLF $= \frac{1}{2\xi}$ and, if the DLF is differentiated with respect to β,

$$\frac{d(\text{DLF})}{d\beta} = 0 \Rightarrow \beta_{\text{peak}} = \sqrt{1 - 2\xi^2}$$

and

$$\frac{\dot{v}}{p} = R \cos(pt - \phi) \quad \text{where} \quad \phi = \tan^{-1}\left(\frac{C_2}{C_1}\right) \tag{4.18}$$

The general solution given by Equation (4.15), which includes the effect of damping, can be used to obtain a better understanding of the resonance phenomenon. As mentioned previously, the first term represents the transient response to the applied loading. This term will dampen out quickly and therefore is of little interest. The second term is the steady-state response, the amplitude of which can be represented as

$$\rho = \frac{p_0}{k} \left[(1 - \beta^2)^2 + (2\xi\beta)^2\right]^{-1/2} \tag{4.19}$$

The dynamic amplification factor for resonance can be determined as being inversely proportional to the damping ratio:

$$R_{\beta=1} = \frac{p}{p_0/k} = \left[(1 - \beta^2)^2 + (2\xi\beta)^2\right]^{-1/2} = \frac{1}{2\xi} \tag{4.20}$$

This result indicates that the dynamic amplification factor for a damped system approaches a limit that is inversely proportional to the damping ratio, 2ξ.

Combining Equations (4.11) and (4.12) gives a general solution of the form

$$v(t) = e^{-\xi \omega t} \left(A \sin \omega_D t + B \cos \omega_D t \right) + C_1 \sin pt + C_2 \cos pt \quad (4.21)$$

This form of the general solution is evaluated at resonance, making the following substitutions: $\beta = 1, C_1 = 0, C_2 = -\frac{p_0}{k} \frac{1}{2\xi}$, which results in

$$v(t) = e^{-\xi \omega t} \left(A \sin \omega_D t + B \cos \omega_D t \right) - \frac{p_0}{k} \frac{1}{2\xi}$$

Applying the initial conditions of starting at rest:

$$A = \frac{p_0}{k} \frac{1}{2\sqrt{1 - \xi^2}} \qquad B = \frac{p_0}{k} \frac{1}{2\xi}$$

Now using the following two conditions at resonance, $\omega_D \cong \omega$, $p = \omega$, results in

$$R = \frac{v(t)}{p_0/k} \cong \frac{1}{2\xi} \left(e^{-\xi \omega t} - 1 \right) \cos \omega t \quad (4.22)$$

A plot of the response amplification versus time at resonance for a damped system is shown in Figure 4.5. As in the undamped case, a number of cycles are required for the damped system to reach the limiting value of the response ratio, and these can be estimated as $1/\xi$, as shown in Figure 4.6. The frequency-response curves for different values of damping are summarized in Figure 4.7.

Resonance can occur in buildings that are subjected to base accelerations having a frequency that is close to that of the building and

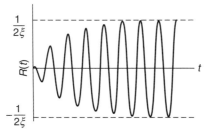

Figure 4.5   Response amplification

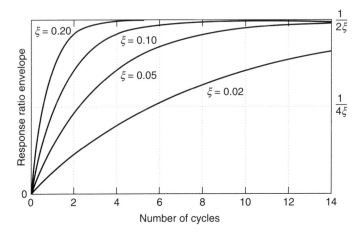

Figure 4.6   Cycles to resonant response

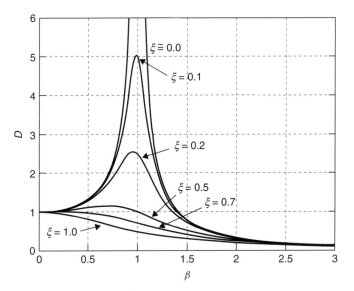

Figure 4.7   Resonant response

having a long duration. It must be recognized that as the response tends to increase, the effective damping will increase, and as cracking of concrete or local yielding of steel occurs, the period of the structure will shift. Both of these actions in the building will tend to reduce the maximum response.

Example 4.3(M)   Use MATLAB to plot the displacement magnification factor, R_d, as a function of β for $\xi = 0.001, 0.1, 0.2, 0.7$, and 1.0. If you limit the y-axis (R_d) range of your graph to $0 \le R_d \le 6$, the plot should look like Figure 4.7.

Recalling that

$$R_d = \frac{1}{\sqrt{\left[1 - \beta^2\right]^2 + [2\xi\beta]^2}}$$

we will generate and plot R_d values for $0 \le \beta \le 3$ in increments of 0.01.

First, we set up two arrays to hold values of β and ξ:

```
beta=0.0:0.01:3.0;
zeta=[0.001, 0.1, 0.2, 0.7, 1.0];
```

Then we loop on each value of ξ to calculate one row of a matrix that holds the values of R_d corresponding to the selected ξ value. For simplicity and ease of debugging, we calculate each denominator term separately before combining them. Also note the use of dot products (".*" and "./ ") in vector operations.

```
for i=1:5
    z=zeta(i);
    denom1= (1-beta.*beta).^2;
    denom2=(2*z*beta).^2;
denom=sqrt(denom1+denom2);
    Rd(i,:)=1./denom;
end
```

Finally, we plot and format the results obtained for R_d as a function of β. Here we have done a little more work to create appropriate legends and line types for this plot. The entire script is as follows:

```
clear all
clc
%
beta=0.0:0.01:3.0;
%
zeta=[0.001, 0.1, 0.2, 0.7, 1.0];
%
for i=1:5
    z=zeta(i);
    denom1= (1-beta.*beta).^2;
    denom2=(2*z*beta).^2;
denom=sqrt(denom1+denom2);
    Rd(i,:)=1./denom;
End
```

```
%
% Create figure
figure1 = figure;
% Create axes
axes1 = axes('Parent',figure1);
xlim(axes1,[0 3]); set x limits for the plot
% if you comment out the following statement, you will see the
% huge amplification caused by damping of 0.001 compared to others
ylim(axes1,[0 6]); set y limits for the plot
%
box(axes1,'on');
grid(axes1,'on');
hold(axes1,'all');
%
% Create multiple lines using matrix input to plot
plot1 = plot(beta,Rd,'Parent',axes1,'LineWidth',2,'Color',[0 0 0]);
set(plot1(1),'LineStyle','-.','DisplayName','\zeta = 0.001');
set(plot1(2),'LineStyle',':','DisplayName','\zeta = 0.1');
set(plot1(3),'LineStyle','--','DisplayName','\zeta = 0.2');
set(plot1(4),'DisplayName','\zeta =0.7');
set(plot1(5),'LineWidth',3,'DisplayName','\zeta = 1.0');
xlabel('\beta'); % Create xlabel
ylabel('Rd');% Create ylabel
legend(axes1,'show');% Create legend
grid on  % Show the grid
```

Upon execution, the graph shown in Figure 4.8 will be displayed.

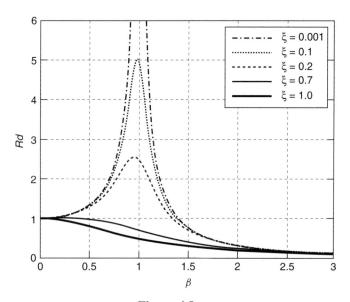

Figure 4.8

4.3 TRIPARTITE LOGARITHMIC PLOT

The displacement amplification factor is given in Equation (4.17). Differentiating this equation with respect to time results in

$$\dot{v} = \frac{p_0}{k} R_d \cos(pt - \phi) p$$

which can be written as

$$\dot{v} = \frac{p_0}{\sqrt{km}} R_v \cos(pt - \phi) \tag{4.23}$$

where $R_v = \left(\dfrac{p}{\omega}\right) R_d = $ the velocity amplification factor

Differentiating Equation (4.20) with respect to time results in

$$\ddot{v} = -\frac{p_0}{k} R_d \frac{p^2}{\omega^2} \sin(pt - \phi) = -\frac{p_0}{m} R_a \sin(pt - \phi) \tag{4.24}$$

where $R_a = \left(\dfrac{p}{\omega}\right)^2 R_d = $ the acceleration amplification factor

From these relationships, it can be seen that

$$\frac{R_a}{p/\omega} = R_v = \left(\frac{p}{\omega}\right) R_d$$

This representation leads to a three-way (tripartite) chart that allows R_d, R_v, and R_a to be plotted versus p/ω, as shown in Figure 4.9.

4.4 EVALUATION OF DAMPING

1. *Free-vibration decay:* It has been shown in Equation (3.35) that the damping for a lightly damped system can be expressed in terms of the logarithmic decrement and the frequency ratio, β. If $\beta \cong 1$, the percentage of critical damping can be estimated as

$$\xi \cong \frac{\delta}{2\pi} \tag{4.25}$$

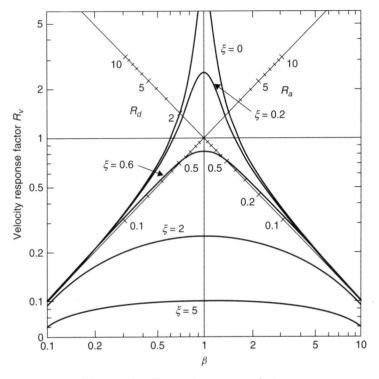

Figure 4.9 Harmonic response factors

2. *Resonate amplification:* In Equation (4.17), it was shown that, at resonance, the response amplification factor approaches a limit that is inversely proportional to the damping ratio:

$$D = \frac{p_{\max}}{p_0} \cong \frac{1}{2\xi}$$

which results in

$$\xi \cong \frac{p_0}{2p_{\max}} \tag{4.26}$$

3. *Half-power method:* Using this method, we can determine the damping from the frequencies at which the response is reduced to $\frac{1}{\sqrt{2}} p_{\max}$. This implies the power input is half the input at resonance. For this reason, the reduced amplitude is often referred to as the *half-power amplitude*. This can be written as

$$\frac{p_{\max}}{\sqrt{2}} = \frac{p_0}{2\xi\sqrt{2}} = p_0 \left[\frac{1}{(1 - \beta^2)^2 + (2\xi\beta)^2} \right]^{-1/2}$$

Squaring both sides and solving for the frequency ratio, we get the following result:

$$\xi = \frac{\Delta \beta}{2} = \frac{f_1 + f_2}{2} \tag{4.27}$$

4. This result indicates that the response amplification factor for a damped system approaches a limit that is inversely proportional to the damping ratio, 2ξ.

Example 4.4(M)   A dance studio is located on the second floor of a building. Because some dance participants complained of excessive floor vibration while dancing exercises were being conducted, the vibration characteristics of the dance floor were measured. The results indicated that the dance floor had a natural frequency of 3.0 Hz in the vertical direction and a damping ratio of $\xi = 0.02$. The excitation caused by dancing on this floor can be approximated by a harmonic function in the form of $p_0 \sin 2\pi f_0 t$, where $1.5 \leq f_0 \leq 3.0$ Hz. The vertical static deflection of the dance floor under the weight of the participants is 0.15 in. The acceptable limit of floor acceleration for this project is set to 1.0 in/sec^2. Use MATLAB to determine the maximum floor acceleration caused by dancing. What level of additional damping is necessary to satisfy the acceleration limit?

From Equation (4.24):

$$\ddot{v} = -\frac{p_0}{k} R_d \frac{p^2}{\omega^2} \sin(pt - \phi)$$

where

$$R_d = \frac{1}{\sqrt{\left[1 - \beta^2\right]^2 + [2\xi\beta]^2}}$$

Therefore, the maximum value of $\ddot{v}$, given $\dfrac{p_0}{k} = 0.015$, is obtained from

$$\ddot{v}_{max} = \left| 0.15 \left(R_d \frac{p^2}{\omega^2} \right) \right|$$

where

$$p = 2\pi f_0 \quad \text{and} \quad \beta = \frac{p}{\omega} = \frac{2\pi f_0}{2\pi (3.5 \text{ Hz})} = \frac{f_0}{3.0}$$

We modify the MATLAB script of Example 4.3(M) to calculate and plot $\ddot{v}_{max}$ as a function of f_0 in the range of interest for values of ξ varying from 0.02 (existing floor) to 0.10 in increments of 0.02.

Here is the modified script:

```
clear all
clc
%
f0=1.5:0.01:3.0;
zeta=[0.02, 0.04, 0.06, 0.08, 0.10];
%
beta=f0./3.0;
%
%
for i=1:5
    z=zeta(i);
    denom1= (1-beta.*beta).^2;
    denom2=(2*z*beta).^2;
denom=sqrt(denom1+denom2);
    Rd(i,:)=1./denom;
    maxAcc(i,:)=0.15.*Rd(i,:).*beta.^2;
end

% Create figure
figure1 = figure;

% Create axes
axes1 = axes('Parent',figure1);
xlim(axes1,[1.5 3]);

box(axes1,'on');
grid(axes1,'on');
hold(axes1,'all');

% Create multiple lines using matrix input to plot
plot1 = plot(f0,maxAcc,'Parent',axes1,'LineWidth',2,
'Color',[0 0 0]);
set(plot1(1),'LineWidth',3,'DisplayName','\zeta = 0.02');
set(plot1(2),'LineStyle',':','DisplayName','\zeta = 0.04');
set(plot1(3),'LineStyle','--','DisplayName','\zeta = 0.06');
set(plot1(4),'DisplayName','\zeta =0.08');
set(plot1(5),'LineStyle','-.','DisplayName','\zeta = 0.10');

% Create xlabel
xlabel('Dancing Frequency, Hz.');

% Create ylabel
ylabel('Max. Acceleration');

% Create legend
legend(axes1,'show');

grid on
```

Upon execution, the graph shown in Figure 4.10 will be displayed from which it can be seen that the existing dance floor acceleration exceeds the 1.0 in/sec^2 limit for excitation frequencies close to 3.0 Hz. Increasing the damping by 6 percent ($\xi = 0.08$) would rectify the problem.

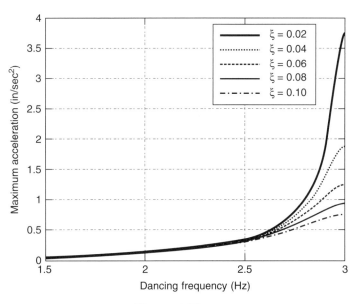

Figure 4.10

4.5 SEISMIC ACCELEROMETERS AND DISPLACEMENT METERS (SEISMOGRAPHS)

Two important dynamic measurement devices used in earthquake engineering are the accelerometer, which measures acceleration, and the seismograph, which measures displacement. These instruments consist primarily of a damped sprung mass that is mounted in a housing that is attached to the surface where the motion is to be measured. The response of the mass is measured relative to that of the housing, as shown in Figure 4.11.

The equation of motion for this system is

$$m\ddot{v} + c\dot{v} + kv = -m\ddot{v}_g(t) = p_{\text{eff}}(t) \qquad (4.28)$$

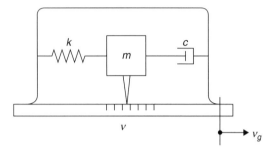

Figure 4.11 Seismic pickup

Because any arbitrary base motion can be resolved into sinusoidal components by use of a Fourier series or a Fourier transform, a simple sinusoidal base motion can be used to indicate the behavior of the instrument under a more general base motion. Therefore, if we let $\ddot{v}_g = \ddot{v}_0 \sin pt$, the solution to the modified equation of motion becomes

$$m\ddot{v} + c\dot{v} + kv = m\ddot{v}_0 \sin pt \qquad (4.29)$$

and the general solution is given as

$$v(t) = \frac{m\ddot{v}_0}{k} \frac{1}{\left[(1 - \beta^2)^2 + (2\xi\beta)^2\right]^{1/2}} \sin(pt - \theta) \qquad (4.30)$$

Introducing the dynamic amplification factor (DAF), we get

$$v_{max} = \frac{m\ddot{v}_0}{k} [\text{DAF}] \qquad (4.31)$$

A plot of the DAF versus the frequency ratio, β, is shown in Figure 4.12. Here it can be seen that, for high damping ($\xi = 0.7$) and a high natural frequency such that $\beta < 0.6$, the DAF is near unity and the instrument is an accelerometer. Hence, the instrument characteristics are high natural frequency and high damping:

$$v_g = v_{g0} \sin pt$$

which leads to

$$\ddot{v}_g = -p^2 v_{g0} \sin pt$$

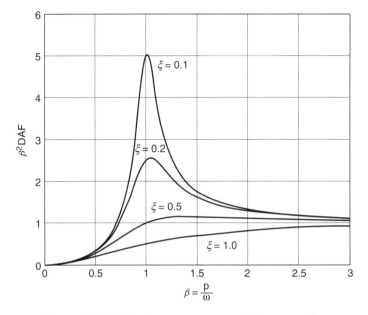

Figure 4.12   Displacement response (seismograph)

and

$$v_{max} = \frac{mp^2 v_{g0}}{k} \, [DAF] = \beta^2 \, [DAF] \, v_{g0} \qquad (4.32)$$

A plot of β versus $\beta^2 DAF$ is shown in Figure 4.12. This figure indicates that, for a damping of approximately 0.5 and a frequency ratio greater than unity, the $\beta^2 DAF$ has a nearly constant value of 1.0, thereby making the measured response proportional to the displacement and the instrument a seismograph.

PROBLEMS

Problem 4.1

A vertical cantilever is constructed using a 3 in × 3 in × 3/16 in steel tube. The tube is 60 in long and supports a 2000 lb weight attached at the tip, as shown in Figure 4.13. This system is subjected to a sinusoidal force at the tip, acting horizontally in one of the planes of symmetry. The force has an amplitude of 250 lb and oscillates at 3 cycles per second. Neglecting the damping, find the maximum steady-state bending stress in the cantilever. Treat the attached weight as a point mass and neglect the weight of the tube. The properties of the tube are as follows: $A = 2.02$ in^2, $S = 1.73$ in^3, $I = 2.6$ in^4, and $E = 29{,}000$ ksi.

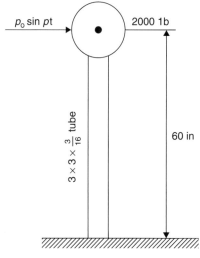

Figure 4.13

Problem 4.2(M)

Solve Problem 4.1 using MATLAB.

Problem 4.3

The water tank shown in Figure 4.14 is subjected to a base excitation that has an acceleration amplitude of $0.1g$ and is idealized as simple harmonic motion with a frequency of 1 Hz. Determine the motion of the tower relative to the motion of the foundation. What is the maximum shear force at the foundation if the damping is assumed to be 5 percent critical?

Problem 4.4(M)

Solve Problem 4.3 using MATLAB.

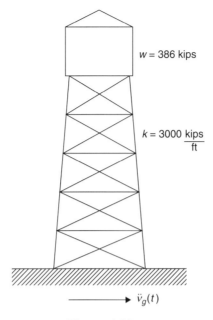

$w = 386$ kips

$k = 3000 \dfrac{\text{kips}}{\text{ft}}$

$\ddot{v}_g(t)$

Figure 4.14

Problem 4.5

A rigid beam 6 ft long and weighing 10 lb/ft rests on a fulcrum 2 ft from one end and is supported by a weightless spring of stiffness 720 lb/ft at one end, as shown in Figure 4.15. At the other end is a viscous damper,

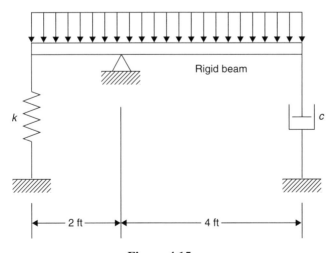

Rigid beam

k

c

2 ft

4 ft

Figure 4.15

$c = 0.30$ lb/ft/sec. The beam is driven by a force that is uniformly distributed along the length of the beam and that has a sinusoidal variation with time, having an amplitude of 12 lb/ft and a frequency of 1.5 cycles per second. Find the amplitude of the steady-state displacement.

Problem 4.6(M)

Solve Problem 4.5 using MATLAB.

Problem 4.7

For the single-story, single-bay frame shown in Figure 4.16, assume the girder is rigid relative to the column and determine the following:

a. The steady-state amplitude for the horizontal motion assuming that the damping is 5 percent of critical damping
b. The horizontal displacement at time $t = 5.0$ sec
c. The transmissibility
d. The maximum flexural stress in the column

The properties of the W10 $\times$ 33 wide-flange section are as follows:
$A = 9.71$ in^2, $I_x = 170$ in^4, $Z_x = 38.8$ in^3, and $S_x = 35.0$ in^3.

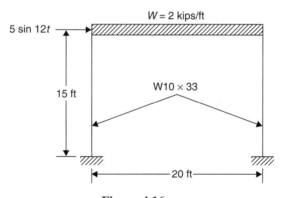

Figure 4.16

Problem 4.8(M)

Solve Problem 4.7 using MATLAB.

CHAPTER 5

RESPONSE TO IMPULSE LOADS

The simplest case of forced vibration occurs when a system that is initially at rest is subjected to a driving force that abruptly jumps from zero to a constant value (zero rise time). This type of loading is often referred to as a *block pulse*. For these dynamic systems, damping is of less importance in controlling the maximum response than for harmonic loads. This is because the maximum response will be reached in a very short time on the first response cycle. This type of response leaves little time for the damping mechanism to absorb much of the energy from the structure. For this reason, emphasis will be placed on the response of an undamped system to impulse loads.

The simplest type of block pulse is one that comes on the system with zero rise time and stays on the system, as shown in Figure 5.1.

The equation of motion for $t > 0$ has the form

$$m\ddot{v}(t) + kv(t) = p_0 \tag{5.1}$$

Note that, although the force is constant, the response is dynamic. As in the previous case for harmonic loads, the general solution to this equation consists of a complementary solution and a particular solution. The complementary solution has the form

$$v_c(t) = A \sin \omega t + B \cos \omega t$$

and the particular solution is

$$v_p(t) = \frac{p_0}{k}$$

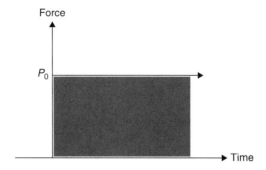

Figure 5.1 Step driving force

The combined solution is

$$v(t) = A \sin \omega t + B \cos \omega t + \frac{p_0}{k} \quad (5.2)$$

Applying the starting-at-rest initial conditions, we have

$$v(t) = \frac{p_0}{k}(1 - \cos \omega t) \quad (5.3)$$

where $\dfrac{p_0}{k}$ = the static deflection

$(1 - \cos \omega t)$ = the dynamic load factor (DLF)

Note that the maximum DLF occurs when $\omega t = \pi$ or when $t = T/2$ and has a value of 2.0. The displacement response of the undamped system for a step driving force is shown in Figure 5.2.

Example 5.1 The single-bay frame shown in Figure 5.3 is discretized as having a rigid beam with flexible columns. It is loaded by a step pulse, as shown, having a force of 55 kips, which is applied to the structure with

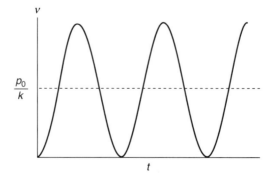

Figure 5.2 Undamped response to a step driving force

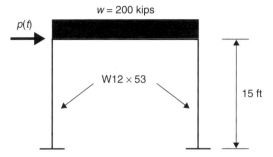

Figure 5.3

a zero rise time. Determine the maximum displacement, the maximum base shear, and the maximum lateral resistance of the two columns.

$W \, 12 \times 53 : I = 425, \, S = 70.6, \, Z = 77.9$, A572 Grade 50

$$k = 2 \left(\frac{12EI}{l^3} \right) = \frac{2(12)(29{,}000)(425)}{[(15)(12)]^3} = 50.72 \, \frac{\text{kips}}{\text{in}}$$

$$m = \frac{w}{g} = \frac{200}{386.4} = 0.518 \, \frac{\text{kip-sec}^2}{\text{in}}$$

$$\omega = \sqrt{\frac{k}{m}} = \sqrt{\frac{50.72}{0.578}} = 9.9 \, \frac{\text{rad}}{\text{sec}} \qquad T = \frac{2\pi}{\omega} = 0.63 \, \text{sec}$$

$$v_{\text{static}} = \frac{P_0}{k} = \frac{55}{50.72} = 1.084 \, \text{in} \qquad v_{\text{dyn}} = \text{DLF} \left(v_{\text{static}} \right)$$

$$= 2(1.084) = 2.17 \, \text{in}$$

$$Q_{\text{max}} = k(v_{\text{max}}) = 50.72(2.17) = 110 \, \text{kips}$$

Now let the moment capacity of the columns increase to the full plastic moment capacity:

$$M_p = F_y Z = 36(77.9) = 2804 \, \text{in-kips}$$

Considering the column fixed at top and bottom and taking moments about the point of inflection, we obtain the following for the lateral force capacity:

$$R = V = 2 \left(\frac{M_p}{h/2} \right) = \frac{4M_p}{h} = \frac{4(2804)}{15(12)} = 62.3 \, \text{kips} \leq 110 \, \text{kips}$$

Because the maximum elastic resistance is less than the demand, the structure will be pushed into the inelastic range and will have a permanent displacement when the load is removed.

5.1 RECTANGULAR PULSE

Now consider a block pulse in which the force jumps from zero to p_0 at time $t = 0$ and then drops back to zero at time $t = t_d$, as shown in Figure 5.4.

The solution for the equation of motion for the time interval $0 < t < t_d$ is

$$v(t) = \frac{p_0}{k}(1 - \cos \omega t) \tag{5.4}$$

For $t > t_d$, the pulse ends, the system is in free vibration, and the solution becomes

$$v(t) = \frac{\dot{v}(t_d)}{\omega} \sin \omega (t - t_d) + v(t_d) \cos \omega (t - t_d) \tag{5.5}$$

The initial conditions for the free-vibration phase are obtained at the end of the forced-vibration phase. By applying these conditions to the above, the solution for the free-vibration phase takes the form

$$v(t) = \frac{p_0}{k} \left[(1 - \cos \omega t_d) \cos \omega (t - t_d) + \sin \omega t_d \sin \omega (t - t_d) \right] \tag{5.6}$$

If the pulse duration, t_d, is greater than $T/2$, the maximum response will be twice the static response and will occur in the first interval, $0 < t < t_d$. If $t_d < T/2$, the maximum response will occur in the free-vibration phase, $t > t_d$, and will be given by

$$v_{max} = \sqrt{\left(\frac{\dot{v}(t_d)}{\omega} \right)^2 + (v(t_d))^2} \tag{5.7}$$

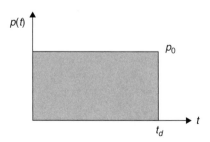

Figure 5.4 Rectangular block pulse

Substituting for $\frac{\dot{v}(t_d)}{\omega}$ and $v(t_d)$ and using the trig identity $\sin^2 \alpha = \frac{1}{2}(1 - \cos 2\alpha)$, we obtain

$$D = \frac{v_{max}}{p_0/k} = 2 \sin \frac{\pi t_d}{T} \qquad \frac{t_d}{T} < \frac{1}{2}$$

Figure 5.5 shows the solution for three different values of the duration of loading, t_d. The values for the duration considered are T, $T/2$, and $T/5$, where the only change in the driving force is the pulse length. However, it can be seen that there is a significant change in the nature of the response. This indicates that the time characteristics of the driving force may be as important as or even more so than its magnitude.

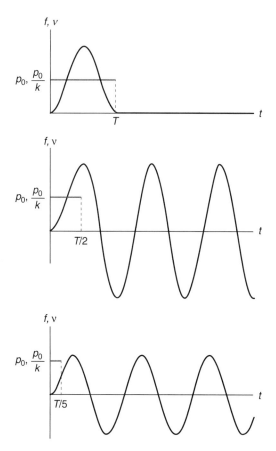

Figure 5.5   Effect of duration of loading

Example 5.2(M)   A system with a natural period of $T = 1.0$ sec is subjected to a rectangular pulse. Use MATLAB to compute and plot the ratio of maximum displacement response to static displacement during the first 5 sec of the system response for the following values of pulse duration: (a) $t_d = T/4$, (b) $t_d = T/2$, (c) $t_d = T$, and (d) $t_d = 2T$.

The desired ratio is obtained by setting the static displacement, $\dfrac{p_0}{k}$, to unity in Equations (5.4) and (5.6). We have MATLAB prompt us for the value of t_d/T :

```
r= input('Enter the value for td/T ratio:');
```

We create a time vector spanning from 0 to 5 sec in 1000 equal steps:

```
n= 1000;
t=linspace(0,5*T,n);
```

and we calculate the value of response at each instant of time:

```
for i=1:n
    if t(i) <= td
        v(i)=(p0/k)*(1-cos(omega*t(i)));
    else
        arg1=(1-cos(omega*td))*cos(omega*(t(i)-td));
        arg2=sin(omega*td)*sin(omega*(t(i)-td));
        v(i) = (p0/k)*(arg1+arg2);
    end
end
```

Finally, we plot the figure. The complete script is as follows:

```
clear all
clc
%
r= input('Enter the value for td/T ratio:');
T=1.0;
omega = 2*pi/T;
td=T*r;
p0=1;
k=1;
%
n= 1000;
t=linspace(0,5*T,n);
%
for i=1:n
    if t(i) <= td
        v(i)=(p0/k)*(1-cos(omega*t(i)));
    else
        arg1=(1-cos(omega*td))*cos(omega*(t(i)-td));
        arg2=sin(omega*td)*sin(omega*(t(i)-td));
```

```
      v(i) = (p0/k)*(arg1+arg2);
   end
end
% Create figure
figure1 = figure;
% Create axes
axes1 = axes('Parent',figure1);
xlim(axes1,[0 5]);
ylim(axes1,[-2.5 2.5]);
box(axes1,'on');
grid(axes1,'on');
hold(axes1,'all');
% Create plot
plot(t,v,'LineWidth',2,'Color',[0 0 0]);
xlabel('Time (sec)');
ylabel('Displacement or Amplification');
```

Executing the script four times, in response to the prompts with values of 0.25, 0.50, 1.0, and 2.0, will result in computation of displacements and display of the graphs shown in Figure 5.6.

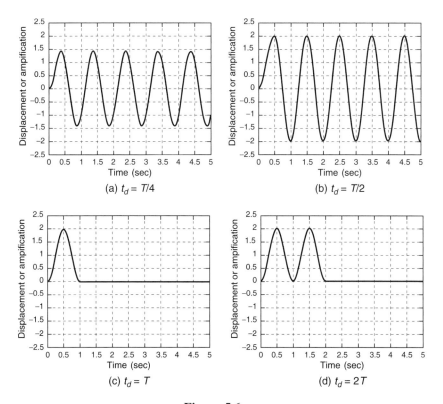

Figure 5.6

5.2 DAMPED RECTANGULAR PULSE

Now consider an ideal rectangular pulse load applied to a damped system. The equation of motion can be written as

$$m\ddot{v} + c\dot{v} + kv = p_0 \tag{5.8}$$

Starting at rest gives the initial conditions $v_0 = \dot{v}_0 = 0$, and the particular solution becomes $v_p = p_0/k$. Combining the particular solution with the homogeneous solution results in

$$v = \frac{p_0}{k} + e^{-\xi\omega t}(A\cos\omega_D t + B\sin\omega_D t) \tag{5.9}$$

Applying the initial conditions as before allows determination of the constants A and B as

$$v_0 = 0 \Rightarrow A = \frac{p_0}{k} \quad \text{and} \quad \dot{v}_0 = 0 \Rightarrow B = -\left[\frac{\xi}{\sqrt{1-\xi^2}}\right]\frac{p_0}{k}$$

Making these substitutions into Equation (5.9) results in the following:

$$v(t) = \frac{p_0}{k}\left\{1 - e^{-\xi\omega t}\left(\cos\omega_D t + \frac{\xi}{\sqrt{1-\xi^2}}\sin\omega_D t\right)\right\} \tag{5.10}$$

This oscillation is shown in Figure 5.7, where it can be seen that the system now oscillates about the static displacement position and the displacement approaches the static displacement as the time becomes large. As before, the dynamic load factor is $\text{DLF} = v(t)_{\text{max}}/v_{\text{static}}$.

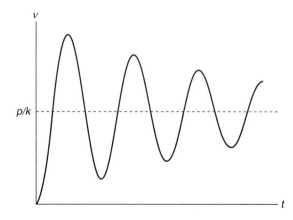

Figure 5.7 Damped response to rectangular pulse

5.3 TRIANGULAR PULSE

Dynamic loading as a result of air blast is often approximated as a decreasing triangular pulse of short duration, as shown in Figure 5.8.

As before, the total solution is considered as a combination of the homogeneous solution and the particular solution as follows:

$$v(t) = v_h + v_p$$

where

$$v_h = A \sin \omega t + B \cos \omega t$$

and

$$v_p = \frac{p_0}{k} \left(1 - \frac{t}{t_1} \right)$$

Using the starting-at-rest initial conditions:

$$v_0 = 0 \Rightarrow B = -\frac{p_0}{k} \qquad \text{and} \qquad \dot{v}_0 = 0 \Rightarrow A = \frac{p_0}{k} \frac{1}{\omega t_1}$$

The general solution then becomes

$$v(t) = \frac{p_0}{k} \left(\frac{\sin \omega t}{\omega t_1} - \cos \omega t - \frac{t}{t_1} + 1 \right) \tag{5.11}$$

Evaluating this equation and its first derivative at the end of phase I ($t = t_1$) results in the following:

$$v(t_1) = \frac{p_0}{k} (\sin \omega t_1 - \cos \omega t_1)$$

$$\dot{v}(t_1) = \frac{p_0 \omega}{k} \left(\frac{\cos \omega t_1}{\omega t_1} + \sin \omega t_1 - \frac{1}{\omega t_1} \right)$$

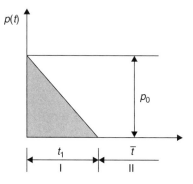

Figure 5.8 Triangular pulse

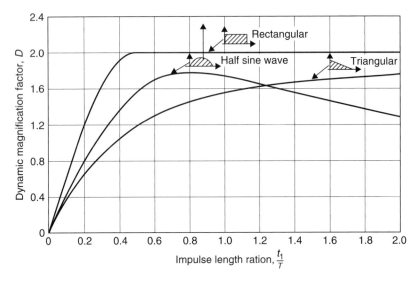

Figure 5.9 Displacement shock spectra for impulse load[1]

These values become the initial conditions for the free-vibration phase of the oscillation:

$$v(\bar{t}) = \frac{\dot{v}(t_1)}{\omega} \left(\sin \omega \bar{t} + v(t_1) \cos \omega \bar{t} \right) \qquad (5.12)$$

The maximum values of these response functions are determined by evaluating them for the times at which the zero-velocity condition is achieved. For very short duration loading, the maximum response occurs during the free-vibration phase. For longer-duration loading, it occurs during the loading phase.

The envelope of maximum values is referred to as the *shock spectra*, a typical representation of which is shown in Figure 5.9. The shock spectra depend only on the shape of the pulse and the duration of the loading relative to the natural period of the structure.

Example 5.3 Determine the maximum elastic deflection and stress for the simple beam shown in Figure 5.10.

---

[1] US Army Corps of Engineers, *Design of Structures to Resist the Effects of Atomic Weapons*, Manual EM 1110-345-415, 1957.

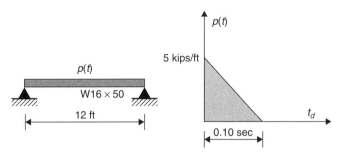

Figure 5.10

The generalized properties of the given system are determined as

$$m^* = \frac{(0.800)(12)}{386.4} = 0.025 \; \frac{\text{kip-sec}^2}{\text{in}}$$

$$k^* = \frac{49.15EI}{L^3} = \frac{49.15(29{,}000)(659)}{(144)^3} = 314.6 \; \frac{\text{kips}}{\text{in}}$$

where m^* and k^* are the generalized mass and stiffness, respectively. Approximating the deflected shape by a sine curve (see Example 2.5), we get the following:

$$p^* = \int_0^L w(x,t)\phi(x)dx = w\int_0^L \sin\frac{\pi x}{L}dx = 0.64W$$

$$m^* = \int_0^L \mu(x)\phi^2(x)dx = \mu\int_0^L \sin^2\frac{\pi x}{L}dx = \frac{\mu L}{2} = \frac{M}{2} = 0.0124$$

The circular frequency is determined as

$$\omega = \sqrt{\frac{k^*}{m^*}} = \sqrt{\frac{314.6}{0.0124}} = 159.3 \; \frac{\text{rad}}{\text{sec}}$$

and the period becomes

$$T = \frac{2\pi}{\omega} = 0.0394 \text{ sec}$$

Then

$$\frac{t_d}{T} = \frac{0.10}{0.039} = 2.54$$

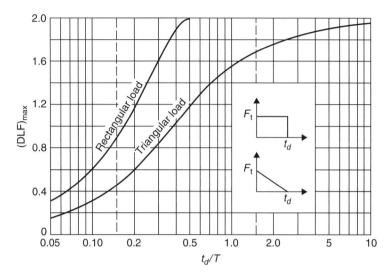

Figure 5.11

and, from the shock spectra shown in Figure 5.11,[2] the DLF = 1.8.

$$p^* = 0.64W = 0.64 \times 5 \times 12 = 38.4 \text{ kips}$$

$$y_{\text{static}} = \frac{p^*}{k^*} = \frac{38.4}{314.6} = 0.122 \text{ in}$$

$$y_{\text{dyn}} = \text{DLF}(y_{\text{static}}) = 1.8(0.122) = 0.22 \text{ in}$$

$$y_{\text{max}} = y_{\text{static}} + y_{\text{dyn}} = 0.34 \text{ in}$$

5.4 APPROXIMATE ANALYSIS FOR SHORT-DURATION IMPULSE LOAD

For longer-duration loads, $t_d/T > 1$, the DLF depends on the rate of increase of the load. For example: for step loading, DLF = 2.0; for sinusoidal loading, DLF = 1.7; and for gradual loading, DLF = 1.0. For short-duration loads, $t_d/T < 0.25$, the maximum response depends on the applied impulse; $I = \int_0^{t_d} p(t)dt$. Because the applied load has a short

[2]US Army Corps of Engineers, *Design of Structures to Resist the Effects of Atomic Weapons*, Manual EM 1110-345-415, 1957.

duration, the damping will be neglected, and the equation of motion has the form

$$m\ddot{v} + kv = p(t) \tag{5.13}$$

On rearranging,

$$m\ddot{v} = p(t) - kv \tag{5.14}$$

Integrating both sides with respect to t, we get

$$m\Delta\dot{v} = \int_0^{t_d} (p(t) - kv(t))dt \tag{5.15}$$

Note that $v(t_d) \approx t_d^2$ and $\Delta\dot{v} \approx t_d$ and therefore $v(t_d)$ is small for small t_d and can be neglected. With this simplification, Equation (5.11) becomes

$$\Delta\dot{v} = \frac{1}{m} \int_0^{t_d} p(t)dt \tag{5.16}$$

For $t > t_d$,

$$v(\bar{t}) = \frac{\dot{v}(t_d)}{\omega} \sin\omega\bar{t} + v(t_d)\cos\omega\bar{t} \tag{5.17}$$

For $\bar{t} = t - t_d$, however, $v(t_d) = $ small $\Rightarrow$ neglect and $\dot{v}(t_d) = \Delta\dot{v}$, which results in

$$v(\bar{t}) = \frac{1}{m\omega}\left(\int_0^{t_d} p(t)dt\right)\sin\omega\bar{t} \tag{5.18}$$

Finally,

$$v_{\max} \cong \frac{1}{m\omega}\int_0^{t_d} p(t)dt \tag{5.19}$$

where $\int_0^{t_d} p(t)dt = $ the area of a short-duration pulse

Example 5.4   Consider the frame of Example 5.1 subjected to a short-duration pulse, as shown in Figure 5.12.

$$k = 50.72\text{kips/in} \quad m = 0.518 \text{ kip-sec}^2/\text{in} \quad \omega = \sqrt{\frac{k}{m}} = 9.9 \frac{\text{rad}}{\text{sec}}$$

$$T = \frac{2\pi}{\omega} = 0.63 \text{ sec}$$

$$\frac{t_d}{T} = \frac{0.05}{0.63} = 0.08 \leq 0.25$$

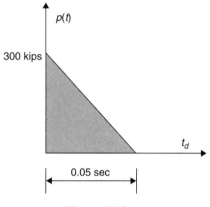

Figure 5.12

Therefore, use approximate analysis:

$$v_{max} = \frac{\int_0^{0.05} p(t)\,dt}{m\omega} = \frac{\frac{1}{2}(300)(0.05)}{0.518(9.9)} = 1.46 \text{ in}$$

$$Q_{max} = kv_{max} = 50.72(1.46) = 74.1 \text{ kips} > R$$

$$= 62.3 \text{ kips (the structure does not remain elastic)}$$

Example 5.5(M) Use MATLAB to directly solve the differential equation of motion for the simple beam of Example 5.3 subjected to the triangular pulse shown for that example.

From Example 5.3,

$$m^* = 0.0124 \text{ kip-sec}^2/\text{in} \quad k* = 314.6 \text{ kips/in}$$

$$\omega = 159.3 \text{ rad/sec} \quad p* = 38.4 \text{ kips}$$

We will modify the script developed for Example 3.8(M) to consider the external force applied to the system.

Equation (5.8) can be rewritten as

$$\ddot{v} + 2\xi\omega\dot{v} + \omega^2 v = \frac{p^*}{m^*} \quad \text{or} \quad \ddot{v} + \omega^2 v = \frac{p^*}{m^*}$$

With the same variable substitution we used for Example 3.8(M):

$$v_1 = v$$
$$v_2 = \dot{v}$$

Converting our equation to the following two first-order equations:

$$\frac{dv_1}{dt} = v_2$$
$$\frac{dv_2}{dt} = -\omega^2 v_1 + \frac{p^*}{m^*}$$

where, for the triangular pulse of Example 5.3,

$$p^* = \begin{cases} \dfrac{38.4 - 384t}{0.0124} & t \le 0.10 \\ 0 & t > 0.10 \end{cases}$$

We have to modify the function DLSDOF that we used in Example 3.8(M) to incorporate the existence of external force. We call this new function DLSDOFP and denote $\dfrac{p^*}{m^*}$ as P. We also simplify the script, taking advantage of the fact that $\xi = 0$.

```
function v = DLSDOFP (t, v)
% define the forcing function
%
m=0.0124;
%
if t<=0.1
    P=(-384*t+38.4)/m;
else
    P=0;
end
%
omega=159.3
%
v= [v(2); -omega*omega*v(1)+P];
```

We divide the time between 0 and 1 sec into 10,000 equal time steps. Therefore, the time array is defined as

```
tspan=linspace(0,1,10000);
```

Next, we call the ode45 solver and plot the displacements:

```
[t, v] = ode45(@DLSDOFP, tspan, [0 0]', []);
plot(t,v(:,1));
```

The complete script is as follows:

```
tspan=linspace(0,1,10000);
[t, v] = ode45(@DLSDOFP, tspan, [0 0]', []);
plot(t,v(:,1));
% Create xlabel
xlabel('t','FontSize',24,'FontName','Times New Roman',
'FontAngle','italic');
% Create ylabel
ylabel('v','FontSize',24,'FontName','Times New Roman',
'FontAngle','italic');
%
% Display maximum value of displacement response
vmax=max(v(:,1))
%
function v = DLSDOFP(t, v)
% define the forcing function
m=0.0124;
if t<=0.1
    P=(-384*t+38.4)/m;
else
    P=0;
end
omega=159.3;
%
v= [v(2); -omega*omega*v(1)+P];
```

Once executed, the graph shown in Figure 5.13 will be displayed, and the maximum dynamic displacement value will be the same as that calculated in Example 5.3.

Recalling that dynamic displacements are calculated with respect to the position of static equilibrium, all we need to do is to add the dynamic and static displacements to obtain the total displacement:

$$y_{static} = \frac{p^*}{k^*} = \frac{38.4}{314.6} = 0.122 \text{ in}$$

$$y_{dyn} = 0.22 \text{ in}$$

$$y_{max} = y_{static} + y_{dyn} = 0.34 \text{ in}$$

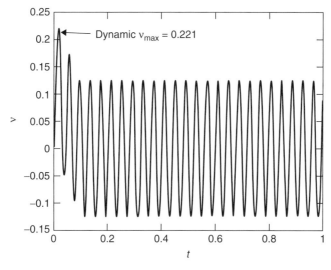

Figure 5.13

PROBLEMS

Problem 5.1

The fixed-end beam shown in Figure 5.14 is subjected to a 70 ksi impulse load, which has a zero rise time and a 0.1 sec duration. The dead load of 1 kip/ft includes an allowance for the beam weight, and it can be assumed that the beam is braced laterally so that the possibility of buckling can be neglected. It is required that the bending stress in the beam not exceed 30 ksi. Does a W24 × 94 wide-flange section ($I = 2700$ in^4, $S = 222$ in^3) meet this criterion?

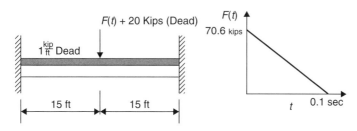

Figure 5.14

Problem 5.2(M)

Solve Problem 5.1 using MATLAB.

Problem 5.3

The single-bay frame shown in Figure 5.15 is subjected to an impulse load having the shape of a half sine wave and a peak load of 500 kips. Assume the beam is rigid relative to the columns and estimate the maximum stress in the columns. Divide the impulse into four segments of 0.1 sec each and integrate numerically using Simpson's rule. The properties of the columns are as follows: $I = 1000$ in^4 and $S = 143$ in^3.

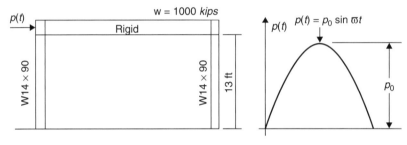

Figure 5.15

Problem 5.4(M)

Solve Problem 5.3 using MATLAB and its ODE45 function without dividing the impulse into segments or using Simpson's rule. Compare your result to that obtained in Problem 5.3. Which result is more accurate and why?

Problem 5.5

The simply supported beam shown in Figure 5.16 is subjected to a uniformly distributed pulse load with a rectangular load-time function. The beam supports a dead load of 1.0 kip/ft in addition to its own weight. Estimate the maximum bending stress in the beam and the maximum deflection, assuming it is braced in the lateral direction. The $W 21 \times 68$ wide-flange section has the following properties: $I = 1480$ in^4 and $S = 140$ in^3.

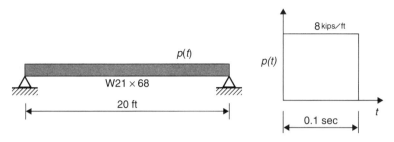

Figure 5.16

Problem 5.6(M)

Solve Problem 5.5 using MATLAB.

CHAPTER 6

RESPONSE TO ARBITRARY DYNAMIC LOADING

6.1 DUHAMEL INTEGRAL

The method described in Chapter 5 for determining the approximate response of a short-duration impulse load can be used as the basis for developing a method for evaluating the response to an arbitrary dynamic load. First, consider an undamped oscillator subjected to a short-duration rectangular pulse having an amplitude of $P(t)$ and a duration of $d\tau$ that ends at time t_1, as shown in Figure 6.1.

From Equation (5.18), the resulting incremental displacement is determined as

$$dv(t) = \frac{P(\tau)d\tau}{m\omega} \sin \omega(t - \tau) \tag{6.1}$$

The total displacement can then be determined by summing all of the incremental displacements in the time interval:

$$v(t) = \int_0^t dv(t) = \int_0^t \frac{P(\tau)}{m\omega} \sin \omega(t - \tau)d\tau \tag{6.2}$$

Equation (6.2) is generally known as the *Duhamel integral* for an undamped elastic system. It may be used to evaluate the response of

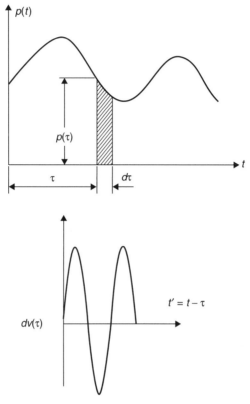

Figure 6.1 Undamped Duhamel integral response (F. Naeim, *The Seismic Design Handbook*, 2nd ed. (Dordrecht, Netherlands: Springer, 2001), reproduced with kind permission from Springer Science+Business Media B.V.)

an undamped single-degree-of-freedom (SDOF) system to any form of dynamic loading. The technique, however, has two major limitations:

1. For arbitrary loadings, evaluation of the integral will have to be done using numerical methods.
2. The solution applies only to elastic response because the principle of superposition is used in the development of the method.

It must also be noted that the Duhamel integral is the particular solution for a system starting at rest. For conditions other than starting at rest, the free-vibration response must be added to this solution, resulting in

$$v(t) = v_h + v_p = \frac{\dot{v}_0}{\omega} \sin \omega t + v_0 \cos \omega t + \int_0^t \frac{P(\tau)}{m\omega} \sin \omega (t - \tau) d\tau$$

$$(6.3)$$

6.2 NUMERICAL FORMULATION OF THE EQUATION OF MOTION

For most arbitrary loadings, the use of numerical methods will be required. Therefore, it is generally expedient to go directly to a numerical solution of the equation of motion that can be used for both linear and nonlinear systems. This section will consider SDOF systems, although the procedures discussed can be readily adapted to multiple-degree-of-freedom (MDOF) systems, as will be shown in Chapters 7 and 8. The applied force and the stiffness are functions of time, whereas the mass and damping are constant. The damping coefficient may also be considered to be a function of time; however, general practice is to determine the damping characteristics for an elastic system and then keep these constant for the complete time history. In the nonlinear range, the primary mechanism for energy dissipation is through inelastic deformation, and this is accounted for by the hysteretic behavior of the restoring force.

Although several integration schemes are available in the literature, a powerful method for doing this is the use of what is generally called the *step-by-step integration method*. In this procedure, the time-dependent equation of dynamic equilibrium is divided into a number of small time increments, and equilibrium must be satisfied at every increment of time. By considering the time at the end of a short time step, the equation of motion can be written as

$$f_i(t + \Delta t) + f_d(t + \Delta t) + f_s(t + \Delta t) = p(t + \Delta t) \qquad (6.4)$$

where $f_i(t + \Delta t) = m\ddot{v}(t + \Delta t)$

$f_d(t + \Delta t) = c\dot{v}(t + \Delta t)$

The restoring force can be written in incremental form as

$$f_s = \sum_{i=1}^{n} k_i(t)\Delta v_i(t) = r(t) + k(t)\Delta v(t) \qquad (6.5)$$

where $\Delta v(t) = v(t + \Delta t) - v(t) =$ the incremental displacement during the current time step

$r(t) = \sum_{i=1}^{n-1} k_i(t)\Delta v_i(t) =$ the elastic restoring force at the beginning of the time interval

It should be noted that the incremental stiffness for a nonlinear system is generally defined as the tangent stiffness at the beginning of the time interval. Making these substitutions into Equation (6.2) results in the following form of the equation of dynamic equilibrium:

$$m\ddot{v}(t + \Delta t) + c\dot{v}(t + \Delta t) + \sum k_i \Delta v_i = P(t + \Delta t) \qquad (6.6)$$

6.3 NUMERICAL INTEGRATION METHODS

Depending on the assumed variation of the acceleration during a small time step, the method may also be referred to as either the *linear acceleration method* or the *constant acceleration method*. If the acceleration is assumed to be constant during the time interval, the equations for the constant variation of the acceleration, the linear variation of the velocity, and the quadratic variation of the displacement are indicated in Figure 6.2. Evaluating the expression for the velocity and displacement at the end of the time interval leads to the following two expressions for velocity and displacement:

$$\dot{v}(t + \Delta t) = \dot{v}(t) + \ddot{v}(t + \Delta t)\frac{\Delta t}{2} + \ddot{v}(t)\frac{\Delta t}{2} \qquad (6.7)$$

$$v(t + \Delta t) = v(t) + \dot{v}(t)\Delta t + \ddot{v}(t + \Delta t)\frac{\Delta t^2}{4} + \ddot{v}(t)\frac{\Delta t^2}{4} \qquad (6.8)$$

Solving Equation (6.8) for the acceleration at the end of the time interval results in

$$\ddot{v}(t + \Delta t) = \frac{4}{\Delta t^2}\Delta v - \frac{4}{\Delta t}\dot{v}(t) - \ddot{v}(t) \qquad (6.9)$$

This can be written as

$$\ddot{v}(t + \Delta t) = \frac{4}{\Delta t^2}\Delta v + A(t) \qquad (6.10)$$

where $\Delta v = v(t + \Delta t) - v(t)$

$$A(t) = -\frac{4}{\Delta t}\dot{v}(t) - \ddot{v}(t)$$

Equation (6.9) expresses the acceleration at the end of the time interval as a function of the incremental displacement and the acceleration and velocity at the beginning of the time interval. Substituting Equation (6.9)

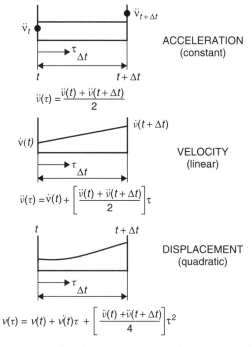

$$\ddot{v}(\tau) = \frac{\ddot{v}(t) + \ddot{v}(t + \Delta t)}{2}$$

$$\dot{v}(\tau) = \dot{v}(t) + \left[\frac{\ddot{v}(t) + \ddot{v}(t + \Delta t)}{2} \right] \tau$$

$$v(\tau) = v(t) + \dot{v}(t)\tau + \left[\frac{\ddot{v}(t) + \ddot{v}(t + \Delta t)}{4} \right] \tau^2$$

Figure 6.2   Increment motion (constant acceleration)   (F. Naeim, *The Seismic Design Handbook*, 2nd ed. (Dordrecht, Netherlands: Springer, 2001), reproduced with kind permission from Springer Science+Business Media B.V.)

into Equation (6.7) results in the following expression for the velocity at the end of the time increment:

$$\dot{v}(t + \Delta t) = \frac{2}{\Delta t} \Delta v - \dot{v}(t) \tag{6.11}$$

which can be written as

$$\dot{v}(t + \Delta t) = \frac{2}{\Delta t} \Delta v + B(t) \tag{6.12}$$

where

$$B(t) = -\dot{v}(t)$$

For the SDOF system, it is convenient to express the damping as a linear function of the mass as

$$c = \alpha m = \lambda C_{\text{cr}} = \lambda 2 m \omega \tag{6.13}$$

Use of this equation allows the proportionality factor, α, to be expressed as $\alpha = 2\lambda\omega$. Substituting Equations (6.10) and (6.12) into Equation (6.6) results in the following form of the equation of motion:

$$m\left[\frac{4}{\Delta t^2}\Delta v + A(t)\right] + \alpha m\left[\frac{2}{\Delta t}\Delta v + B(t)\right] + \sum k_i \Delta v_i = P(t + \Delta t)$$
(6.14)

Moving terms containing the response conditions at the beginning of the time interval to the right side of the equation results in the following so-called pseudostatic form of the equation of motion:

$$\bar{k}_t(\Delta v) = \bar{p}(t + \Delta t) \tag{6.15}$$

where

$$\bar{k}_t = \frac{4m}{\Delta t^2} + \frac{2\alpha m}{\Delta t} + k_t$$

and

$$\bar{p}(t + \Delta t) = p(t + \Delta t) - r(t) - m[A(t) + \alpha B(t)]$$

where

$$r(t) = \sum_{\tau=0}^{\tau=t} k_\tau \Delta v_\tau$$

and $r(t)$ is the resistance at the beginning of the time interval. The incremental displacement during the time increment can be written as

$$\Delta v = \frac{\bar{p}}{\bar{k}_t} \tag{6.16}$$

The displacement, velocity, and acceleration at the end of the time increment can then be determined as

$$v(t + \Delta t) = v(t) + \Delta v \tag{6.17}$$

$$\dot{v}(t + \Delta t) = \frac{2}{\Delta t} + B(t) \tag{6.18}$$

$$\ddot{v}(t + \Delta t) = \frac{4}{\Delta t^2} + A(t) \tag{6.19}$$

These values become the initial conditions for the next time increment, and the procedure is repeated. If the acceleration during a small time step is assumed to have a *linear variation*, the following three expressions for the displacement, velocity, and acceleration at the end of the time interval can be determined in a similar manner:

$$v(t + \Delta t) = v(t) + \dot{v}(t)\Delta t + \ddot{v}(t)\frac{\Delta t^2}{3} + \ddot{v}(t + \Delta t)\frac{\Delta t^2}{6} \tag{6.20}$$

$$\dot{v}(t + \Delta t) = \dot{v}(t) + \ddot{v}(t)\Delta t + \ddot{v}(t + \Delta t)\frac{\Delta t}{2} - \ddot{v}(t)\frac{\Delta t}{2} \tag{6.21}$$

$$\ddot{v}(t + \Delta t) = \frac{6}{\Delta t^2}\Delta v - \frac{6}{\Delta t}\dot{v}(t) - 2\ddot{v}(t) \tag{6.22}$$

The equations for acceleration and velocity at the end of the time step can also be written as

$$\ddot{v}(t + \Delta t) = \frac{6}{\Delta t^2}(\Delta v) + A(t) \tag{6.23}$$

$$\dot{v}(t + \Delta t) = \frac{3}{\Delta t}(\Delta v) + B(t) \tag{6.24}$$

where

$$A(t) = -\frac{6}{\Delta t}\dot{v}(t) - 2\ddot{v}(t)$$

$$B(t) = -2\dot{v}(t) - \frac{\Delta t}{2} - \ddot{v}(t)$$

6.4   NEWMARK'S NUMERICAL METHOD

Newmark[1] suggested a numerical procedure for structural dynamics that is similar to the step-by-step method discussed previously. This integration scheme has the following general form:

$$\dot{v}(t + \Delta t) = \dot{v}(t) + (1 - \gamma)\ddot{v}(\Delta t) + \gamma\ddot{v}(t + \Delta t)\Delta t \tag{6.25}$$

$$v(t + \Delta t) = v(t) + \dot{v}(t)\Delta t + \left(\frac{1}{2} - \beta\right)\ddot{v}(t)\Delta t^2 + \beta\ddot{v}(t + \Delta t)\Delta t^2 \tag{6.26}$$

If $\gamma = \frac{1}{2}$ and $\beta = \frac{1}{4}$, the Newmark method becomes the same as the constant acceleration method. If $\gamma = \frac{1}{2}$ and $\beta = \frac{1}{6}$, the linear acceleration during the time increment is obtained.

[1]N. M. Newmark, "A Method of Computation for Structural Dynamics," *Trans. ASCE*, Vol. 127, 1962.

Based on his studies, Newmark reached the following conclusions regarding the proposed integration method:

1. If $\gamma \neq \dfrac{1}{2}$, the integration procedure will introduce a spurious damping into the system even without real damping in the problem.
2. If $\gamma = 0$, negative damping results, and this induces self-excited vibration solely as a result of the numerical integration procedure.
3. If $\gamma > 1$, a positive damping is introduced that will reduce the response.

In order to carry out numerical integration with the Newmark method, a step-by-step procedure would be useful. The following is one such procedure.[2]

1. Perform the initial calculations:

    a. $\ddot{v}(0) = \dfrac{P(0) - c\dot{v}(0) - kv(0)}{m}$

    b. $\bar{k} = k + \dfrac{\gamma}{\beta \Delta t}c + \dfrac{1}{\beta(\Delta t)^2}m$

    c. $A = \dfrac{1}{\beta \Delta t}m + \dfrac{\gamma}{\beta}c$

    d. $B = \dfrac{1}{2\beta}m + \Delta t\left(\dfrac{\gamma}{2\beta} - 1\right)c$

2. Calculate for each time step i:

    a. $\Delta \bar{P}_i = \Delta P_i + A\dot{v}_i + B\ddot{v}_i$

    b. $\Delta v_i = \dfrac{\Delta \bar{P}_i}{\bar{k}}$

    c. $\Delta \dot{v}_i = \dfrac{\gamma}{\beta \Delta t}\Delta v_i - \dfrac{\gamma}{\beta}\dot{v}_i + \Delta t\left(1 - \dfrac{\gamma}{2\beta}\right)\ddot{v}_i$

    d. $\Delta \ddot{v}_i = \dfrac{1}{\beta(\Delta t)^2}\Delta v_i - \dfrac{1}{\beta \Delta t}\dot{v}_i - \dfrac{1}{2\beta}\ddot{v}_i$

    e. $v_{i+1} = v_i + \Delta v_i, \dot{v}_{i+1} = \dot{v}_i + \Delta \dot{v}_i, \ddot{v}_{i+1} = \ddot{v}_i + \Delta \ddot{v}_i$

3. Repeat step 2, replacing i with $i + 1$ and continue.

Example 6.1(M)   A tower can be modeled as an SDOF system with a weight of 386.4 kips, a stiffness of 39.48 kips/in, and 10 percent critical damping. The tower is subjected to a dynamic load of $P(t) =$

---

[2]A. K. Chopra, *Dynamics of Structures*, 3rd ed. (Upper Saddle River, NJ: Pearson Education, 2007).

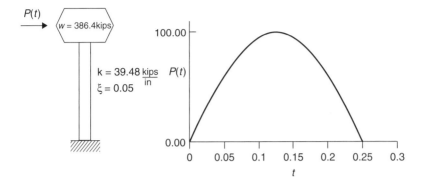

Figure 6.3

$100 \sin(4\pi t)$, where $0 \le t \le 0.25$ sec, as shown in Figure 6.3. Use MAT-LAB with a time step of $\Delta t = 0.01$ sec to calculate the displacement of the tower during its first 6 sec of response.

a. Use MATLAB's built-in integration function and ODE45.
b. Use Newmark's method with $\gamma = \frac{1}{2}$ and $\beta = \frac{1}{4}$ (constant acceleration method).
c. Use Newmark's method with $\gamma = \frac{1}{2}$ and $\beta = \frac{1}{6}$ (linear acceleration method).
d. Compare the results obtained in parts (a)–(c).

$$m = \frac{w}{g} = \frac{386.4}{386.4} = 1.00 \qquad \omega = \sqrt{\frac{k}{m}} = \sqrt{\frac{39.48}{1.00}} = 6.28$$

a. Use MATLAB's built-in integration function and ode45.

All we need to do is to modify the script developed for Example 5.5(M) to include damping and the new definition of external load. Recalling

$$\ddot{v} + 2\xi\omega\dot{v} + \omega^2 v = \frac{P}{m}$$

and using the same variable substitution we used before:

$$v_1 = v$$
$$v_2 = \dot{v}$$

Converting our equation to the following two first-order equations:

$$\frac{dv_1}{dt} = v_2$$

$$\frac{dv_2}{dt} = -\omega^2 v_1 - 2\xi \omega v_2 + \frac{P}{m}$$

where

$$P = \begin{cases} 100\sin(4\pi t) & t \le 0.25 \\ 0 & t > 0.25 \end{cases}$$

We have to modify the function DLSDOFP that we used in Example 5.5(M) accordingly. We call this new function DLSDOFP2 and denote P/m as P.

```
function v = DLSDOFP2 (t, v)
% define the forcing function
%
m=1;
k=39.48;
zeta =0.10
omega=sqrt(k/m);
%
if t<=0.25
    P=100*sin(4*pi()*t)/m;
else
    P=0;
end
%
%
v= [v(2); -omega*omega*v(1)-2*zeta*omega*v(2)+P];
```

You can experiment with various time steps to get a feel for how many of them are necessary to obtain accurate results. For this problem, probably any number above 200 time steps will provide you with good results. Because we wanted a very smooth curve for our plots, we have gone overboard and have divided the time between 0 and 6 sec into 10,000 equal time steps. Even with these many time steps, MATLAB solves the problem in an instant. The time array is defined as follows:

```
tspan=linspace(0,6,10000);
```

Next, we call the ode45 solver and plot the displacements:

```
[t, v] = ode45(@DLSDOFP2, tspan, [0 0]', []);
plot(t,v(:,1));
```

The complete script is as follows:

```
tspan=linspace(0,6,10000);
[t, v] = ode45(@DLSDOFP2, tspan, [0 0]', []);
plot(t,v(:,1));
% Create xlabel
xlabel('t','FontSize',24,'FontName','Times New Roman',
'FontAngle','italic');
% Create ylabel
ylabel('v','FontSize',24,'FontName','Times New Roman',
'FontAngle','italic');
%
% Display maximum value of displacement response
vmax=max(v(:,1))
%
function v = DLSDOFP2 (t, v)
% define the forcing function
%
m=1;
k=39.48;
zeta =0.10
omega=sqrt(k/m);
%
if t<=0.25
    P=100*sin(4*pi()*t)/m;
else
    P=0;
end
%
%
v= [v(2); -omega*omega*v(1)-2*zeta*omega*v(2)+P];
```

Once executed, the graph shown in Figure 6.4 will be displayed, and the maximum dynamic displacement value will be shown on the MATLAB workspace as

```
vmax =
   2.0652.
```

b. Use Newmark's method with $\gamma = \frac{1}{2}$ and $\beta = \frac{1}{4}$.

First, define the time array, time step, mass, stiffness, and damping coefficient as well as γ and β:

```
t=linspace(0,6,10000);
Dt= t(2)-t(1);
m=1;
k=39.48;
zeta =0.10;
omega=sqrt(k/m);
c=2*m*omega*zeta;
```

```
%
gamma= 1/2;
beta = 1/4;
```

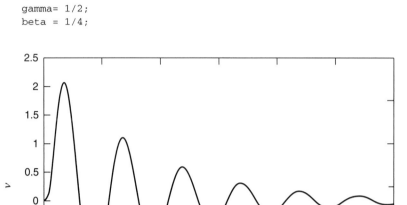

Figure 6.4

Next, define the forcing function, P, and using the MATLAB function diff, an array containing changes in P during each time step:

```
for i = 1:length(t)
    if t(i)<=0.25
        P(i)=100*sin(4*pi()*t(i));
    else
        P(i)=0;
    end
end
DP = diff(P)
```

Now we can implement the step-by-step procedure:

```
%
% Initial calculations
%
v(1) =0;
vdot(1) = 0;
vdotdot(1) = (P(1) - c*vdot(1) -k*v(1))/m;
```

```
kbar = k + gamma*c/(beta*Dt) + m/(beta*Dt*Dt);
A = m/(beta*Dt) + gamma*c/beta;
B= m/(2*beta) +Dt*c*((0.5*gamma/beta)-1)*c;
%
% Loop over each time step
%
for i = 1:(length(t)-1)
    DPbar = DP(i) + A*vdot(i)+B*vdotdot(i);
    Dv = DPbar/kbar;
    Dvdot = gamma*Dv/(beta*Dt) - gamma*vdot(i)/beta +
    Dt*vdotdot(i)*(1-0.5*gamma/beta);
    Dvdotdot= Dv/(beta*Dt*Dt) - vdot(i)/(beta*Dt)
    -vdotdot(i)/(2*beta);
    v(i+1) = v(i) + Dv;
    vdot(i+1) = vdot(i) + Dvdot;
    vdotdot(i+1) = vdotdot(i) +Dvdotdot;
end
```

Finally, we calculate and display the maximum value and plot the displacement:

```
%
% Find the maximum value of displacement
%
vmax = max(v)
%
% Plot displacement
%
plot(t, v);
% Create xlabel
xlabel('t','FontSize',24,'FontName','Times New Roman',
'FontAngle','italic');
% Create ylabel
ylabel('v','FontSize',24,'FontName','Times New Roman',
'FontAngle','italic');
```

Execution of this script results in $\text{vmax} = 2.0600$ and the displacement graph shown in Figure 6.5.

c. Use Newmark's method with $\gamma = \frac{1}{2}$ and $\beta = \frac{1}{6}$.

All we need to do is to substitute $\gamma = \frac{1}{2}$ and $\beta = \frac{1}{6}$ for gamma and beta in the script developed in part (b), resulting in $\text{vmax} = 2.0974$ and the displacement graph shown in Figure 6.6.

d. Compare the results obtained in parts (a)–(c).

Obviously, the results obtained are virtually identical. The maximum difference among the displacement results obtained is less than 2 percent.

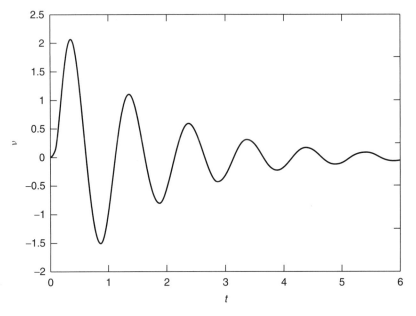

Figure 6.5

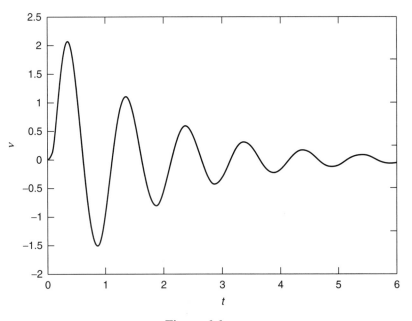

Figure 6.6

PROBLEMS

Problem 6.1(M)

Solve Example 6.1(M) assuming the following values of damping. Compare the results and explain the differences. Recall that we have solved this problem for $\xi = 10\%$.

    a. $\xi = 0\%$
    b. $\xi = 5\%$
    c. $\xi = 20\%$
    d. $\xi = 50\%$
    e. $\xi = 100\%$

Problem 6.2(M)

Solve Example 6.1(M) using the following Δt values. Recall that we have solved this problem using $\Delta t = 0.01$ sec. What is the largest value of Δt that we can use to obtain results within 90 percent accuracy?

    a. $\Delta t = 0.005$ sec
    b. $\Delta t = 0.02$ sec
    c. $\Delta t = 0.05$ sec
    d. $\Delta t = 0.10$ sec

CHAPTER 7

MULTIPLE-DEGREE-OF-FREEDOM SYSTEMS

In the previous chapters, procedures for analyzing structural systems that can be represented by a single degree of freedom and a unique deformed shape have been presented. However, it is not always possible to model the dynamic response accurately in terms of a single displacement coordinate. The generalized single-degree-of-freedom (SDOF) procedure seeks to capture the response of the structure in the fundamental mode of vibration. However, some structures may have more than one possible shape during the dynamic response. This may be a result of one or more of the higher modes of vibration and will require additional displacement coordinates to obtain an accurate representation of the response.

7.1 ELASTIC PROPERTIES

7.1.1 Flexibility

The inverse relationship for the stiffness is defined by the flexibility influence coefficient as

f_{ij} = the deflection at coordinate i caused by a unit load at coordinate j

As an example of the use of flexibility coefficients, consider the simple beam with three vertical displacement degrees of freedom, as shown in Figure 7.1.

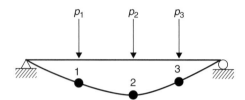

Figure 7.1   Simple beam with static concentrated loads

The determination of the vertical displacements caused by the unit loads is a standard problem in static structural analysis, and any method may be used. When the complete set of influence coefficients has been determined, they can be used to calculate the displacement vector from the applied loads. For example, the displacement at coordinate 1 can be expressed as

$$v_1 = f_{11}p_1 + f_{12}p_2 + f_{13}p_3 \tag{7.1}$$

Similar expressions for the other displacement components lead to the complete set, which can be expressed in matrix form as

$$\begin{Bmatrix} v_1 \\ v_2 \\ v_3 \end{Bmatrix} = \begin{bmatrix} f_{11} & f_{12} & f_{13} \\ f_{21} & f_{22} & f_{23} \\ f_{31} & f_{32} & f_{33} \end{bmatrix} \begin{Bmatrix} p_1 \\ p_2 \\ p_3 \end{Bmatrix} = [F]\{p\} \tag{7.2}$$

where $[F] = $ the flexibility matrix

7.1.2 Stiffness

In a similar manner, the stiffness influence coefficient for the simple beam can be defined as

$k_{ij} = $ the force at coordinate i caused by a unit displacement at coordinate j, all other displacements being zero

In this case, the force at coordinate 1 can be expressed as

$$f_{s1} = k_{11}v_1 + k_{12}v_2 + k_{13}v_3 \tag{7.3}$$

Expressions for the remaining force components lead to the complete set, which can be expressed in matrix form as

$$\begin{Bmatrix} f_{s1} \\ f_{s2} \\ f_{s3} \end{Bmatrix} = \begin{bmatrix} k_{11} & k_{12} & k_{13} \\ k_{21} & k_{22} & k_{23} \\ k_{31} & k_{32} & k_{33} \end{bmatrix} \begin{Bmatrix} v_1 \\ v_2 \\ v_3 \end{Bmatrix} = [K]\{v\} \tag{7.4}$$

where $[K] = $ the stiffness matrix

7.1.3 Inertia

The inertia coefficient is defined as

$m_{ij} = $ the force at coordinate i caused by a unit acceleration at coordinate j

Expressions for the other inertia force components can be expressed in matrix form as

$$\begin{Bmatrix} f_{I1} \\ f_{I2} \\ f_{I3} \end{Bmatrix} = \begin{bmatrix} m_{11} & m_{12} & m_{13} \\ m_{21} & m_{22} & m_{23} \\ m_{31} & m_{32} & m_{33} \end{bmatrix} \begin{Bmatrix} \ddot{v}_1 \\ \ddot{v}_2 \\ \ddot{v}_3 \end{Bmatrix} = [M]\{\ddot{v}\} \tag{7.5}$$

where $[M] = $ the mass matrix

In the case of a lumped-mass system, the mass matrix is diagonal.

7.1.4 Viscous Damping

By analogy, the viscous damping forces can be represented as

$$f_{d1} = c_{11}\dot{v}_1 + c_{12}\dot{v}_2 + c_{13}\dot{v}_3 \tag{7.6}$$

Expressions for the other viscous damping force components can be expressed in matrix form as

$$\begin{Bmatrix} f_{d1} \\ f_{d2} \\ f_{d3} \end{Bmatrix} = \begin{bmatrix} c_{11} & c_{12} & c_{13} \\ c_{21} & c_{22} & c_{23} \\ c_{31} & c_{32} & c_{33} \end{bmatrix} \begin{Bmatrix} \dot{v}_1 \\ \dot{v}_2 \\ \dot{v}_3 \end{Bmatrix} = [C]\{\dot{v}\} \tag{7.7}$$

where $[C] = $ the damping matrix

An initial approximation for obtaining the response of a multiple-degree-of-freedom (MDOF) building system is to consider a typical transverse frame that is part of the lateral force system. The mass of the system is assumed to be concentrated (lumped) at the individual story levels. Each lumped mass includes the entire mass of the area tributary to the frame under consideration, which extends from the midheight of the story below to the midheight of the story above. It is also assumed that there is no rotation of a horizontal section at the floor level and that the beams are rigid relative to the columns. Therefore, a typical building frame will have a single horizontal translational degree of freedom at each story level, and the lateral story stiffness will be due entirely to the stiffness of the columns of the story level in bending. These constraints cause the deflected building to have many of the features of a cantilever beam that is deflected by the action of shear forces. Hence, it is often referred to as the *shear building approximation*. This results in a diagonal matrix of the mass properties in which the translational mass is located on the main diagonal, as shown in Equation (7.8). The resulting stiffness properties for this idealization take a tridiagonal form, as shown in Equation (7.9).

$$\{f_i\} = \begin{bmatrix} m_1 & & & & & \\ & m_2 & & & & \\ & & m_3 & & & \\ & & & \cdot & & \\ & & & & \cdot & \\ & & & & & \cdot \\ & & & & & & m_n \end{bmatrix} \begin{Bmatrix} v_1 \\ v_2 \\ v_3 \\ \cdot \\ \cdot \\ \cdot \\ v_n \end{Bmatrix} \tag{7.8}$$

$$\{f_s\} = \begin{bmatrix} k_1 & -k_2 & & & & \\ -k_2 & k_1 + k_2 & -k_3 & & & \\ & -k_3 & k_2 + k_3 & -k_4 & & \\ & & \cdot & \cdot & \cdot & \\ & & & \cdot & \cdot & \cdot \\ & & & & \cdot & \cdot & -k_{n-1} \\ & & & & & -k_{n-1} & k_{n-1} + k_n \end{bmatrix} \begin{Bmatrix} v_1 \\ v_2 \\ v_3 \\ \cdot \\ \cdot \\ v_{n-1} \\ v_n \end{Bmatrix}$$

$$\tag{7.9}$$

7.2   UNDAMPED FREE VIBRATION

Free vibration implies that the structure is not subjected to any external excitation and that its motion occurs only from the initial conditions of displacement and velocity. This condition is important because the analysis of the structure response in free motion provides important information regarding the dynamic properties of the structure. These properties are the natural frequencies and the corresponding vibration mode shapes.

The equation of motion for free vibration in matrix form can be written as

$$[m]\{\ddot{v}\} + [k]\{v\} = [0] \tag{7.10}$$

If it is assumed that the resulting motion is simple harmonic, the displacement can be expressed as

$$v(t) = A\sin(\omega t + \theta) \tag{7.11}$$

The first and second derivatives become

$$\dot{v}(t) = A\omega\cos(\omega t + \theta) \tag{7.12}$$
$$\ddot{v}(t) = -A\omega^2\sin(\omega t + \theta) = -\omega^2 v(t) \tag{7.13}$$

Substituting these quantities into Equation (7.10) results in the following matrix form for the equation of free vibration:

$$[m]\left\{-\omega^2 v(t)\right\} + [k]\{v(t)\} = [0] \tag{7.14}$$

which can be written as

$$\left[[k] - \omega^2[m]\right]\{v\} = [0] \tag{7.15}$$

The previous matrix equation, also referred to as the *eigenvalue equation*, is a set of homogeneous equilibrium equations, which can have a nontrivial solution only if the determinant of the coefficient matrix is zero:

$$\left|[k] - \omega^2[m]\right| = [0] \tag{7.16}$$

Expansion of the determinant results in a polynomial of degree N in terms of the circular frequency, ω. Hence, it is sometimes referred to as the *frequency equation*. The N roots of the frequency equation represent the natural frequencies of the N modes of vibration. The mode having the lowest frequency (longest period) is called the *first* or *fundamental mode*.

Example 7.1 Fundamental Periods of Vibration Consider the four-story steel frame of Example 2.5. In the earlier example, the frame was modeled as a generalized SDOF system using the static deflected shape to approximate the first mode shape. The frame will now be modeled as a MDOF system having a horizontal degree of freedom at each story level. The natural frequencies and fundamental periods of the four modes of vibration will be determined in terms of these four degrees of freedom.

Considering the mass of the frame to be lumped at the story levels results in a diagonal mass matrix of the form

$$[M] = \begin{bmatrix} 0.761 & & & \\ & 0.952 & & \\ & & 0.952 & \\ & & & 0.958 \end{bmatrix}$$

Using the shear building approximation for the stiffness results in a tridiagonal stiffness matrix of the form

$$[K] = 500 \begin{bmatrix} 1.175 & -1.175 & 0 & 0 \\ -1.175 & 2.351 & -1.175 & 0 \\ 0 & -1.175 & 2.643 & -1.467 \\ 0 & 0 & -1.467 & 2.451 \end{bmatrix}$$

These two matrices can be combined to form the left side of Equation (7.15):

$$[K] - \omega^2 [M]$$

$$= \begin{bmatrix} 1.175 - 0.761B & -1.175 & 0 & 0 \\ -1.175 & 2.351 - 0.952B & -1.175 & 0 \\ 0 & -1.175 & 2.643 - 0.952B & -1.467 \\ 0 & 0 & -1.467 & 2.451 - 0.958B \end{bmatrix}$$

where

$$B = \frac{\omega^2}{500}$$

Inserting this result into Equation 7.16 and expanding the resulting determinant, we obtain the following fourth-order polynomial, which is also known as the frequency equation, in which the four roots of this polynomial represent the natural frequencies of the four modes of vibration:

$$B^4 - 9.343B^3 + 26.525B^2 - 22.952B + 3.027 = 0$$

$$B = 0.160, 1.253, 3.189, 4.741$$

$$B_1 = \frac{\omega_1^2}{500} = 0.160 \quad \omega_1 = 8.94 \quad T_1 = 0.702$$

$$B_2 = \frac{\omega_2^2}{500} = 1.253 \quad \omega_2 = 25.03 \quad T_2 = 0.251$$

$$B_3 = \frac{\omega_3^2}{500} = 3.189 \quad \omega_3 = 39.93 \quad T_3 = 0.157$$

$$B_4 = \frac{\omega_4^2}{500} = 4.741 \quad \omega_4 = 48.68 \quad T_4 = 0.129$$

When the frequencies are known, the frequency equation can be used to determine the deflected shape (mode shape). The frequency equation for the nth mode becomes

$$[E^n]\{v_n\} = [0] \tag{7.17}$$

where

$$[E^n] = [k] - \omega_n^2[m]$$

Because the frequencies were evaluated from this condition, the equation is satisfied identically and v_n is indeterminate. Therefore, the shape of the mode will be determined in terms of any one coordinate.

For a three-story shear building having one translational degree of freedom per story, the matrix $[E^n]$ can be written as follows with coordinate 1 assumed to be unity:

$$\begin{bmatrix} e_{11} & e_{12} & e_{13} \\ e_{21} & e_{22} & e_{23} \\ e_{31} & e_{32} & e_{33} \end{bmatrix} \begin{Bmatrix} 1 \\ \phi_{21} \\ \phi_{31} \end{Bmatrix} = \begin{Bmatrix} 0 \\ 0 \\ 0 \end{Bmatrix} \tag{7.18}$$

where

$$\phi_{11} = \frac{v_{11}}{v_{11}} = 1 \quad \phi_{21} = \frac{v_{21}}{v_{11}} \quad \phi_{31} = \frac{v_{31}}{v_{11}}$$

The partitioning of the matrix is done to correspond with the yet-unknown displacement amplitudes, which can be expressed symbolically as

$$
\begin{bmatrix} e_{11}^n & E_{10}^n \\ E_{01}^n & E_{00}^n \end{bmatrix} \begin{Bmatrix} 1 \\ v_{0n} \end{Bmatrix} = \begin{Bmatrix} 0 \\ 0 \end{Bmatrix}
\tag{7.19}
$$

which results in

$$
E_{01}^n(1) + E_{00}^n v_{0n} = 0 \quad \text{and} \quad v_{0n} = -(E_{00}^n)^{-1} E_{01}^n
$$

Example 7.2 Normal Modes Determine the deflected shapes (mode shapes) for the four modes of vibration for the four-story steel frame of the previous example:

$$
[K] - \omega^2[M]
$$

$$
= 500 \begin{bmatrix}
1.175 - 0.761B & -1.175 & 0 & 0 \\
-1.175 & 2.351 - 0.952B & -1.175 & 0 \\
0 & -1.175 & 2.643 - 0.952B & -1.467 \\
0 & 0 & -1.467 & 2.451 - 0.958B
\end{bmatrix}
$$

For mode 1, $B = 0.160$, and the characteristic equation becomes

$$
\begin{Bmatrix} \phi_{21} \\ \phi_{31} \\ \phi_{41} \end{Bmatrix} = -\begin{bmatrix} 2.199 & -1.175 & 0 \\ -1.175 & 2.491 & -1.467 \\ 0 & -1.467 & 2.298 \end{bmatrix}^{-1} \begin{Bmatrix} -1.175 \\ 0 \\ 0 \end{Bmatrix} = \begin{Bmatrix} 0.897 \\ 0.678 \\ 0.432 \end{Bmatrix}
$$

For mode 2, $B = 1.253$, and

$$
\begin{Bmatrix} \phi_{22} \\ \phi_{32} \\ \phi_{42} \end{Bmatrix} = -\begin{bmatrix} 1.158 & -1.175 & 0 \\ -1.175 & 1.450 & -1.467 \\ 0 & -1.467 & 1.251 \end{bmatrix}^{-1} \begin{Bmatrix} -1.175 \\ 0 \\ 0 \end{Bmatrix} = \begin{Bmatrix} 0.188 \\ -0.815 \\ -0.956 \end{Bmatrix}
$$

For mode 3, $B = 3.189$, and

$$
\begin{Bmatrix} \phi_{23} \\ \phi_{23} \\ \phi_{43} \end{Bmatrix} = -\begin{bmatrix} 0.685 & -1.175 & 0 \\ -1.175 & -0.393 & -1.467 \\ 0 & -1.467 & -0.604 \end{bmatrix}^{-1} \begin{Bmatrix} -1.175 \\ 0 \\ 0 \end{Bmatrix} = \begin{Bmatrix} -1.048 \\ -0.389 \\ 0.944 \end{Bmatrix}
$$

For mode 4, $B = 4.741$, and

$$\left\{ \begin{array}{c} \phi_{24} \\ \phi_{34} \\ \phi_{44} \end{array} \right\} = - \left[\begin{array}{ccc} -2.162 & -1.175 & 0 \\ -1.175 & -1.871 & -1.467 \\ 0 & -1.467 & -2.091 \end{array} \right]^{-1} \left\{ \begin{array}{c} -1.175 \\ 0 \\ 0 \end{array} \right\} = \left\{ \begin{array}{c} -2.251 \\ 3.143 \\ -2.204 \end{array} \right\}$$

The four mode shapes for this example are shown in Figure 7.2.

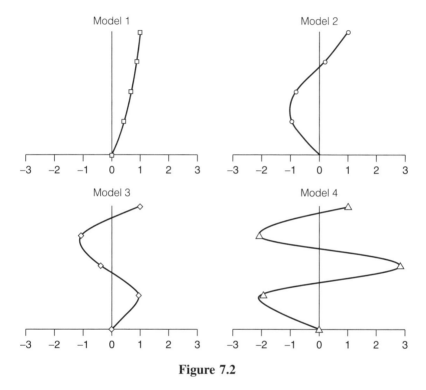

Figure 7.2

Also recall that the SDOF solution for the first mode shape, using the static deflected shape, resulted in the following mode shape and period for the first mode:

$$\{\phi\}^T = [1.000 \quad 0.902 \quad 0.706 \quad 0.470] \qquad T = 0.702$$

Example 7.3(M)   Use MATLAB to calculate the fundamental periods of vibration and mode shapes for the four-story steel frame of the previous two examples. Compare your results to those obtained in Examples 7.1 and 7.2.

As demonstrated in Examples 7.1 and 7.2, hand calculation of natural periods and mode shapes of MDOF systems can be tedious and time-consuming. In contrast, probably nothing in MATLAB is simpler than calculating eigenvalues and eigenvectors and therefore natural frequencies, periods, and mode shapes.

All we need to do is to define the mass and stiffness matrices (K, M) for MATLAB and invoke the eig(K,M) function.

There are several ways to input matrices in MATLAB. One could read them from a file, import them from a spreadsheet, or directly input them into MATLAB. For this example, we will directly input the mass and stiffness matrices as follows:

```
>> M = [0.761 0 0 0; 0 0.952 0 0; 0 0 0.952 0; 0 0 0 0.958]
```

MATLAB echo verifies our input for the mass matrix:

```
M =
    0.7610         0         0         0
         0    0.9520         0         0
         0         0    0.9520         0
         0         0         0    0.9580
>> K=500.*[1.175 -1.175 0 0; -1.175 2.351 -1.175 0; 0 -1.175
           2.643 -1.467; 0 0 -1.467 2.451]
K =
   1.0e+03 *
    0.5875   -0.5875         0         0
   -0.5875    1.1755   -0.5875         0
         0   -0.5875    1.3215   -0.7335
         0         0   -0.7335    1.2255
```

Just to make sure that we have the right values, we divide the stiffness matrix values by 500 (notice that we are using scalar multiplication and division here):

```
>> KT= K./500
KT =
    1.1750   -1.1750         0         0
   -1.1750    2.3510   -1.1750         0
         0   -1.1750    2.6430   -1.4670
         0         0   -1.4670    2.4510
```

Now we invoke the eig(K,M) function to obtain the eigenvalues and eigenvectors:

```
>> [Eigenvectors, Eigenvalues] = eig(K,M)
Eigenvectors =
   -0.6832   -0.6592    0.5930   -0.2473
   -0.6125   -0.1241   -0.6292    0.5139
```

```
   -0.4630      0.5369     -0.2299     -0.7035
   -0.2956      0.6300      0.5644      0.4909
Eigenvalues =
   1.0e+03  *
    0.0799           0           0           0
         0      0.6267           0           0
         0           0      1.5910           0
         0           0           0      2.3765
```

The eigenvalues are ω^2 values. The natural periods (T) are obtained from $T = 2\pi/\omega$. First, we obtain the ω values and then calculate the natural periods:

```
>> omegas=sqrt(Eigenvalues)
omegas =
    8.9385           0           0           0
         0     25.0338           0           0
         0           0     39.8876           0
         0           0           0     48.7497
>> T=2*pi./omegas
T =
    0.7029         Inf         Inf         Inf
       Inf      0.2510         Inf         Inf
       Inf         Inf      0.1575         Inf
       Inf         Inf         Inf      0.1289
```

The diagonal terms of T are the natural periods of the system, and the off-diagonal terms are infinity because we incurred division by zero. To avoid that, we could calculate the natural periods individually:

```
>> T1= 2*pi./omegas(1,1)
T1 = 0.7029
>> T2= 2*pi./omegas(2,2)
T2 = 0.2510
>> T3= 2*pi./omegas(3,3)
T3 = 0.1575
>> T4= 2*pi./omegas(4,4)
T4 = 0.1289
```

Comparing the results obtained using MATLAB to those obtained in Example 7.1, we see that the values are virtually identical:

	T_1	T_2	T_3	T_4
Example 7.1	0.702	0.251	0.157	0.129
MATLAB	0.7029	0.2510	0.1575	0.1289

At a first glance, it may appear that the eigenvectors or mode shapes obtained from MATLAB are different from those obtained in Example 7.2. This is not the case, however, because the mode shapes in Example 7.2 were normalized so that the first value of each mode shape is unity. If we normalize the eigenvectors obtained from MATLAB by dividing each column of the matrix by the value of the first element of that column, we will see that the results are close:

```
>> phi1 =Eigenvectors(:,1)./Eigenvectors(1,1)

phi1 =
    1.0000
    0.8965
    0.6777
    0.4327

>> phi2 =Eigenvectors(:,2)./Eigenvectors(1,2)

phi2 =
    1.0000
    0.1882
   -0.8145
   -0.9557

>> phi3 =Eigenvectors(:,3)./Eigenvectors(1,3)

phi3 =
    1.0000
   -1.0609
   -0.3876
    0.9518

>> phi4 =Eigenvectors(:,4)./Eigenvectors(1,4)

phi4 =
    1.0000
   -2.0784
    2.8453
   -1.9853
```

Compare the matrix of eigenvectors obtained from MATLAB to that obtained in Example 7.2:

$$\phi_{\text{Example 7.2}} = \begin{bmatrix} 1.000 & 1.000 & 1.000 & 1.000 \\ 0.897 & 0.188 & -1.048 & -2.251 \\ 0.678 & -0.815 & -0.389 & 3.143 \\ 0.432 & -0.956 & 0.944 & -2.204 \end{bmatrix}$$

$$
\phi_{\text{MATLAB}} = \begin{bmatrix} 1.000 & 1.000 & 1.000 & 1.000 \\ 0.8965 & 0.1882 & -1.0609 & -2.0784 \\ 0.6777 & -0.8145 & -0.3876 & 2.8453 \\ 0.4327 & -0.9557 & 0.9518 & -1.9853 \end{bmatrix}
$$

7.3 FREE VIBRATION

Consider the four-story building of the previous example. The displaced shape is defined by translational displacement coordinates at four levels. Any displacement vector for this structure can be developed by super-imposing suitable amplitudes of the four modes of vibration. For any modal component, the displacements are given by the mode shape vector multiplied by the modal amplitude:

$$
\{v_n\} = \{\phi_n\} Y_n \tag{7.20}
$$

The total displacement is then obtained as the sum of the modal components

$$
\{v_n\} = \{\phi_n\} Y_n = \phi_1 Y_1 + \phi_2 Y_2 + \cdots\cdots + \phi_N Y_N = \sum_{n=1}^{N} \phi_n Y_n \tag{7.21}
$$

In this equation, the mode shape matrix serves to transform from the generalized coordinates, Y, to the geometric coordinates, $[v]$. The mode amplitude generalized coordinates are called the *normal coordinates* of the structure. Because the mode shape matrix for a system with N degrees of freedom consists of N independent modal vectors, it is nonsingular and can be inverted. If all the modes where known, all the modal vectors could be assembled into a matrix that could be inverted. However, this would require knowing all of the modes and inverting a potentially large matrix. Neither of these alternatives is particularly attractive. A process that gets around this is to use the orthogonality property.

Multiplying both sides of Equation (7.21) by $\phi_n[m]$ gives

$$
\{\phi_n^T\} [m]\{v\} = \{\phi_n^T\} [m] [\phi] [Y] \tag{7.22}
$$

By using the orthogonality property, all terms of this series vanish with the exception of those corresponding to ϕ_n. Therefore, Equation (7.22) reduces to

$$
\{\phi_n^T\} [m]\{v\} = \{\phi_n^T\} [m] \{\phi_n\} Y_n
$$

from which the normal coordinate can be determined as

$$Y_n = \frac{\phi_n^T [m]\{v\}}{\phi_n^T [m]\phi_n} \tag{7.23}$$

Equation (7.23) can be written as

$$Y_0 = [M^*]^{-1} \left[\phi^T\right] [M]\{v_0\}$$

and the free vibration response can be represented as

$$Y_n(t) = \frac{\dot{Y}_{n0}}{\omega_n} \sin \omega_n t + Y_{n0} \cos \omega_n t \tag{7.24}$$

Example 7.4 Free Vibration   Determine the free-vibration response of the four-story steel frame if it is given an initial displacement as follows and a zero initial velocity:

$$\{v_0^T\} = \{9.0 \quad 7.0 \quad 5.0 \quad 3.0\}$$

$$[M] = \begin{bmatrix} 0.761 & 0 & 0 & 0 \\ 0 & 0.952 & 0 & 0 \\ 0 & 0 & 0.952 & 0 \\ 0 & 0 & 0 & 0.958 \end{bmatrix}$$

$$[\phi] = \begin{bmatrix} 1.0 & 1.0 & 1.0 & 1.0 \\ 0.897 & 0.188 & -1.048 & -2.251 \\ 0.678 & -0.815 & -0.389 & 3.143 \\ 0.432 & -0.956 & 0.944 & -2.204 \end{bmatrix}$$

$$[M^*] = [\phi]^T [M][\phi] = \sum M_i \phi_i^2$$

$$M_1^* = 0.761(1)^2 + 0.952(0.897)^2$$
$$+ 0.952(0.678)^2 + 0.958(0.432)^2 = 2.14$$

$$M_2^* = 0.761(1)^2 + 0.952(0.188)^2$$
$$+ 0.952(-0.815)^2 + 0.958(-0.956)^2 = 2.30$$

$$M_3^* = 0.761(1)^2 + 0.952(-1.048)^2$$
$$+ 0.952(-0.389)^2 + 0.958(0.944)^2 = 2.80$$

$$M_4^* = 0.761(1)^2 + 0.952(-2.252)^2$$
$$+ 0.952(3.143)^2 + 0.958(-2.204)^2 = 19.64$$

$$[M^*]^{-1} \quad [\phi]^T \quad [M] \quad \{v_0\} \quad Y(t)$$

$$\begin{bmatrix} 0.47 & 0 & 0 & 0 \\ 0 & 0.43 & 0 & 0 \\ 0 & 0 & 0.36 & 0 \\ 0 & 0 & 0 & 0.05 \end{bmatrix} \begin{bmatrix} 1.0 & 0.90 & 0.68 & 0.43 \\ 1.0 & 0.19 & -0.82 & -0.96 \\ 1.0 & -1.05 & -0.39 & 0.94 \\ 1.0 & -2.25 & 3.14 & -2.2 \end{bmatrix}$$

$$\times \begin{bmatrix} 0.76 & 0 & 0 & 0 \\ 0 & 0.95 & 0 & 0 \\ 0 & 0 & 0.95 & 0 \\ 0 & 0 & 0 & 0.96 \end{bmatrix} \begin{Bmatrix} 9 \\ 7 \\ 5 \\ 3 \end{Bmatrix} = \begin{Bmatrix} 8.11 \cos \omega_1 t \\ 0.68 \cos \omega_2 t \\ 0.22 \cos \omega_3 t \\ 0.11 \cos \omega_4 t \end{Bmatrix}$$

$$v(t) = [\phi]\{Y(t)\} = \{\phi_1\}Y_1 + \{\phi_2\}Y_2 + \{\phi_3\}Y_3 + \{\phi_4\}Y_4$$

$$v_1(t) = 8.11 \cos \omega_1 t + 0.68 \cos \omega_2 t + 0.22 \cos \omega_3 t + 0.11 \cos \omega_4 t$$

$$v_2(t) = 7.27 \cos \omega_1 t + 0.13 \cos \omega_2 t - 0.23 \cos \omega_3 t - 0.25 \cos \omega_4 t$$

$$v_3(t) = 5.50 \cos \omega_1 t - 0.55 \cos \omega_2 t - 0.09 \cos \omega_3 t + 0.35 \cos \omega_4 t$$

$$v_4(t) = 3.50 \cos \omega_1 t - 0.65 \cos \omega_2 t + 0.21 \cos \omega_3 t - 0.23 \cos \omega_4 t$$

Example 7.5(M)   In Example 7.4, time-dependent displacement of each degree of freedom was expressed as a linear combination of a set of coefficients and $\cos \omega_i t$, where i varied from 1 to 4. Use MATLAB to calculate the $\cos \omega_i t$ coefficients of the vector $Y(t)$. Then use MATLAB's Symbolic Math Toolbox to calculate $v(t)$.

We first form $[\phi]$ by assembling vectors of normalized mode shapes:

```
phi(:,1) = phi1; phi(:,2) = phi2; phi(:,3) = phi3; phi(:,4) = phi4;
```

To confirm:

```
>> phi
phi =
    1.0000    1.0000    1.0000    1.0000
    0.8965    0.1882   -1.0609   -2.0784
    0.6777   -0.8145   -0.3876    2.8453
    0.4327   -0.9557    0.9518   -1.9853
```

Then we form $[M^*]$ and define the vector of initial displacements. Notice that "$'$" indicates the transpose of a matrix:

```
>> Mstar=phi'*M*phi
Mstar =
     2.1427    -0.0000    -0.0000     0.0000
    -0.0000     2.3014    -0.0000    -0.0000
    -0.0000    -0.0000     2.8433     0.0000
    -0.0000    -0.0000     0.0000    16.3562
V0=[9.0; 7.0; 5.0; 3.0]
V0 =
     9
     7
     5
     3
```

Now we calculate the desired coefficients by obtaining the inverse of $[M^*]$ and multiplying it by $[\phi]^T$ and $\{v_0\}$:

```
Ycoeff = inv(Mstar)*phi'*M*V0
Ycoeff =
     8.0704
     0.6429
     0.2356
     0.0511
```

Define symbolic variables for $Y, \omega_1, \omega_2, \omega_3$, and ω_4 and represent Y in terms of these variables. Note that we use the MATLAB vpa function to represent the results of symbolic operations in decimals rather than in fractions:

```
>> syms t w1 w2 w3 w4 Y
>> Y= Ycoeff.*[cos(w1*t); cos(w2*t); cos(w3*t); cos(w4*t)]

Y =

 (5044*cos(t*w1))/625
 (6429*cos(t*w2))/10000
  (589*cos(t*w3))/2500
  (511*cos(t*w4))/10000

>> Y= vpa(Y,4)
Y =

   8.07*cos(t*w1)
 0.6429*cos(t*w2)
 0.2356*cos(t*w3)
 0.0511*cos(t*w4)
```

Finally, calculate $v(t) = [\phi] \{Y(t)\}$:

```
>> v= vpa(phi*Y,4)

v =

   8.07*cos(t*w1) + 0.6429*cos(t*w2) + 0.2356*cos(t*w3)
   + 0.0511*cos(t*w4)
   7.235*cos(t*w1) + 0.121*cos(t*w2) - 0.2499*cos(t*w3)
   - 0.1062*cos(t*w4)
   5.469*cos(t*w1) - 0.5236*cos(t*w2) - 0.09132*cos(t*w3)
   + 0.1454*cos(t*w4)
   3.492*cos(t*w1) - 0.6144*cos(t*w2) + 0.2242*cos(t*w3)
   - 0.1014*cos(t*w4)
```

7.4 BETTI'S LAW

A property that is very important in structural dynamics can be obtained by application of Betti's theorem (law). Betti's law states that, for an elastic structure acted upon by two systems of loads and corresponding displacements, the work done by the first system of loads moving through the displacements of the second system is equal to the work done by the second system of loads moving through the displacements produced by the first load system.

Consider two sets of loads, p_a and p_b. Apply loads p_a to the structure and calculate the work done (Figure 7.3):

$$W_a = \frac{1}{2} \sum p_a v_a = \frac{1}{2} \{p_a\}^T \{v_a\} \qquad (7.25)$$

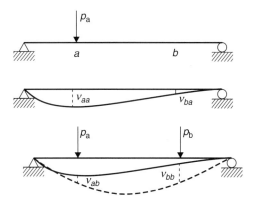

Figure 7.3   Application of Betti's law

Now apply loads p_b and calculate the work done by the application of this set of loads:

$$W_{ab} = \frac{1}{2}\{p_b\}^T\{v_b\} + \{p_a\}\{v_b\} \tag{7.26}$$

The total work done by this sequence of loading can be written as

$$W_{\text{total}} = W_a + W_{ab} = \frac{1}{2}\{p_a\}^T\{v_a\} + \frac{1}{2}\{p_b\}^T\{v_b\} + \{p_a\}^T\{v_b\} \tag{7.27}$$

If the sequence of loading is reversed and loads p_b are applied to the structure:

$$W_b = \frac{1}{2}\{p_b\}^T\{v_b\}$$

Now apply loads p_a and calculate the work done by the application of this set of loads:

$$W_{ba} = \frac{1}{2}\{p_a\}^T\{v_a\} + \{p_b\}^T\{v_a\}$$

The total work done by this sequence of loading is

$$W_{\text{total}} = W_b + W_{ba} = \frac{1}{2}\{p_b\}^T\{v_b\} + \frac{1}{2}\{p_a\}^T\{v_a\} + \{p_b\}^T\{v_a\} \tag{7.28}$$

For an elastic system, the work done is independent of the sequence of load application. Therefore, the total work must be a constant, and the two expressions for the total work must be equal. Equating these two work expressions results in the following, which is an expression of Betti's law:

$$\{p_a\}^T\{v_b\} = \{p_b\}^T\{v_a\}$$

The displacements can be expressed in terms of the flexibility matrix as

$$\{v_b\} = [f]\{p_b\}$$

Substituting this expression into the equation for Betti's law results in

$$\{p_a\}^T[f]\{p_b\} = \{p_b\}^T[f]\{p_a\} \tag{7.29}$$

which indicates that $[f] = [f]^T$; therefore, the flexibility matrix must be symmetric. This is an expression of Maxwell's law of reciprocal deflections.

7.5 ORTHOGONALITY PROPERTIES OF MODE SHAPES

One of the most attractive methods for solving problems having multiple degrees of freedom is through the use of normal modes. In special circumstances, a dynamic system will oscillate at one of the natural frequencies alone while all of the other principal modes are zero. When this happens, it is referred to as a *normal mode of vibration*.

Consider an undamped system in free vibration (Figure 7.4). The equation of motion has the form

$$[m]\{\ddot{v}\} + [k]\{v\} = [0] \tag{7.30}$$

which can be written as

$$[k]\{v\} - \omega^2[m]\{v\} = [0]$$

or

$$[k]\{v\} = \omega^2[m]\{v\} = \text{inertia force}$$

Now apply Betti's law to obtain the following:

$$\sum_i \left(\omega_m^2 \phi_{im} m_i\right)\phi_{in} = \sum_i \left(\omega_n^2 \phi_{in} m_i\right)\phi_{im}$$

$$\left(\omega_m^2 - \omega_n^2\right) \sum \phi_{im} m_i \phi_{in} = 0$$

If

$$\left(\omega_m^2 - \omega_n^2\right) \neq 0$$

then

$$\sum \phi_{im} m_i \phi_{in} = 0$$

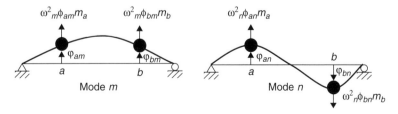

Figure 7.4 Reciprocity of normal modes

In matrix form:

$$\{\phi_m\}^T [M] \{\phi_n\} = 0 \qquad m \neq n \qquad (7.31)$$

Hence, the normal mode shapes are orthogonal with respect to the mass matrix. In a similar manner, consider the stiffness matrix:

$$[k]\{v\} = \omega^2 [m]\{v\}$$

Premultiply both sides by $\{\phi_m\}^T$:

$$\{\phi_m\}^T [k] \{\phi_n\} = \omega^2 \{\phi_m\}^T [m]\{\phi_n\} = 0$$

Therefore,

$$\{\phi_m\}^T [k]\{\phi_n\} = 0 \qquad m \neq n \qquad (7.32)$$

And the normal mode shapes are orthogonal with respect to the stiffness matrix.

7.6 CHANGING COORDINATES (INVERSE TRANSFORMATION)

Consider a structural system in which the deflected shape is defined by translational displacement coordinates at the story levels. For any modal component, v_n, the displacements are given by the mode shape vector, ϕ_n, multiplied by the modal amplitude, Y_n:

$$v_n = \phi_n Y_n$$

These mode amplitude generalized coordinates are called the *normal coordinates* of the structure.

The total displacement is then obtained as the sum of the modal components:

$$v = \phi_1 Y_1 + \phi_2 Y_2 + \cdots + \phi_n Y_n = \sum_{n=1}^{N} \phi_n Y_n \qquad (7.33)$$

Now premultiply Equation (7.33) by $\{\phi_n\}^T [m]$ to obtain

$$\{\phi_n\}^T [m]\{v\} = \{\phi_n\}^T [m] \{\phi_1\} Y_1 + \{\phi_n\}^T [m] \{\phi_2\} Y_2$$
$$+ \cdots + \{\phi_n\}^T [m]\{\phi_n\}Y_N \qquad (7.34)$$

By the orthogonality property of normal modes with respect to the mass, all terms in this series are zero with the exception of those corresponding to ϕ_n. Introducing this term on both sides of the equation gives

$$\{\phi_n\}^T[m]\{\phi\}\{Y\} = \{\phi_n\}^T[m]\{\phi_1\}Y_1 + \{\phi_n\}^T[m]\{\phi_2\}Y_2$$
$$+ \cdots + \{\phi_n\}^T[m]\{\phi_n\}Y_N \qquad (7.35)$$

By orthogonality,

$$\{\phi_n\}^T[m]\{\phi_m\} = 0 \qquad m \neq n$$

$$\{\phi_n\}^T[m]\{v\} = \{\phi_n\}^T[m]\{\phi_n\}Y_n$$

and

$$Y_n = \frac{\{\phi_n\}^T[m]\{v\}}{\{\phi_n\}^T[m]\{\phi_n\}} \qquad (7.36)$$

Example 7.6 Inverse Transformation   Assume that the four-story steel frame of the previous example is given a 1 in displacement at all floors.

$$\begin{bmatrix} 1.0 & 1.0 & 1.0 & 1.0 \\ 0.897 & 0.188 & -1.048 & -2.251 \\ 0.678 & -0.815 & -0.389 & 3.143 \\ 0.432 & -0.956 & 0.944 & -2.204 \end{bmatrix} = [\phi]$$

$$\begin{bmatrix} 0.761 & & & \\ & 0.952 & & \\ & & 0.952 & \\ & & & 0.958 \end{bmatrix} = [m] \qquad \begin{Bmatrix} 1 \\ 1 \\ 1 \\ 1 \end{Bmatrix} = \{v\}$$

Y_1:
$$\{\phi_1\}^T[m]\{v\} = \{1.0\ 0.897\ 0.678\ 0.432\}[m]\{v\} = 2.674$$
$$\{\phi_1\}^T[m]\{\phi_1\} = 2.143 \qquad Y_1 = \frac{2.674}{2.143} = 1.248$$

Y_2:
$$\{\phi_2\}^T[m]\{v\} = \{1.0\ 0.188\ -0.815\ -0.956\}[m]\{v\} = -0.752$$
$$\{\phi_2\}^T[m]\{\phi_2\} = 2.303 \qquad Y_2 = -\frac{0.752}{2.303} = -0.326$$

Y_3:
$$\{\phi_3\}^T[m]\{v\} = \{1.0\ -1.048\ -0.389\ 0.944\}[m]\{v\} = 0.297$$
$$\{\phi_3\}^T[m]\{\phi_3\} = 2.804 \qquad Y_3 = \frac{0.297}{2.804} = 0.106$$

$$Y_4: \quad \{\phi_4\}^T[m]\{v\} = \{1.0 \ -2.251 \ 3.143 \ -2.204\} \, [m]\{v\} = -0.501$$

$$\{\phi_4\}^T[m]\{\phi_4\} = 19.64 \qquad Y_4 = -\frac{0.501}{19.64} = -0.026$$

The resulting mode shapes are shown in Figure 7.5.

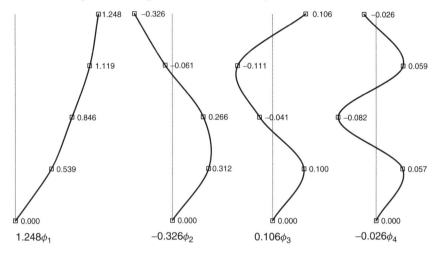

1.248ϕ_1 −0.326ϕ_2 0.106ϕ_3 −0.026ϕ_4

Figure 7.5

Example 7.7(M) Use MATLAB to solve the inverse transformation of Example 7.6 and demonstrate that the summation of modal responses is equal to the unity displacement vector imposed on the structure.

We already have the MATLAB representations of the mass matrix and mode shapes from Example 7.3(M). All we need to do is to define a unity vector, v, and apply Equation (7.36):

```
>> v=[1; 1; 1; 1]
v =
     1
     1
     1
     1
>> Y1=(phi1'*M*v)/(phi1'*M*phi1)
Y1 = 1.2480
>> Y2=(phi2'*M*v)/(phi2'*M*phi2)
Y2 = -0.3262
>> Y3=(phi3'*M*v)/(phi3'*M*phi3)
Y3 = 0.1034
>> Y4=(phi4'*M*v)/(phi4'*M*phi4)
Y4 = -0.0251
```

Now let us confirm that the linear combination of modal responses is equal to the imposed displacement:

```
>> check = Y1*phi1 + Y2*phi2 +Y3*phi3 +Y4*phi4
check =
    1.0000
    1.0000
    1.0000
    1.0000
```

7.7 HOLZER METHOD FOR SHEAR BUILDINGS

The defining feature of a shear building with masses lumped at the story levels is that the shear in each story depends only on the displacement within that story. This type of structural system is said to be *close-coupled*, and the Holzer method is a simple and efficient way of finding one or more of its modes. The method requires that an initial assumption of the frequency (period) be made for the mode to be evaluated. For building structures, this can be done for the fundamental mode by using one of the approximate formulas specified in typical building codes. Necessary conditions for the use of this method are that the inertia matrix is diagonal and the stiffness matrix is tridiagonal.

If the oscillation of the structure is assumed to be sinusoidal, the displacement can be taken as

$$v_n = v_0 \sin \omega t \tag{7.37}$$

and successive differentiation with respect to time results in the following expression for the acceleration at the nth floor level:

$$\ddot{v}_n = \omega^2 v_0 \sin \omega t = \omega^2 v_n \tag{7.38}$$

The inertia force at the top floor becomes

$$F_n = m_n \omega^2 v_n F_n = m_n \ddot{v}_n = m_n \omega^2 v_n \tag{7.39}$$

which is equal to the shear in the top story, V_n. Knowing the stiffness of the top story, we can calculate the displacement within the story as

$$\Delta v_n = \frac{V_n}{K_n} \tag{7.40}$$

The displacement of the story below can then be determined as $v_{n-1} = v_n - \Delta v_n$. This process is repeated until the displacement at the base is finally obtained:

$$v_0 = v_1 - \Delta v_1 \tag{7.41}$$

If the base displacement is zero, ω is a natural frequency, and the displaced shape is a mode shape. The base displacement amplitude is a function of ω^2 and the zeros of that function give the natural frequencies of the system.

Newton's method is a technique for finding the roots of a polynomial representation of any continuous and differentiable function. Once two calculations have been completed, Newton's method can be used to obtain an improved estimate for the third iteration. The procedure can be summarized as follows:

$$r_{i+1} = r_i - \frac{f(r_i)}{f'(r_i)} \tag{7.42}$$

where $f'(\omega^2)$ can be approximated as

$$f'(\omega^2) \cong \frac{f(\omega^2 + \delta\omega^2) - f(\omega^2)}{\delta\omega^2} \tag{7.43}$$

$$\omega_{n+1}^2 = \omega_n^2 - \frac{f(\omega^2)\,\delta\omega^2}{f(\omega^2 + \delta\omega^2) - f(\omega^2)} \tag{7.44}$$

From the results of two trials, a good estimate of the true frequency can be obtained by linear extrapolation:

$$\left(\frac{\Delta\omega^2}{\Delta v_0}\right)_{1-2} = \left(\frac{\Delta\omega^2}{\Delta v_0}\right)_{2-3} \tag{7.45}$$

The Holzer process for shear buildings has the advantage that any mode can be calculated independently of the other modes. Other procedures require that the modes be calculated in sequence, and any error in one mode propagates to all subsequent modes. It should be noted, however, that Newton's method of iteration may not necessarily converge to the frequency of interest. Hence, the Holzer method is better suited to an interactive approach in which the user provides the initial approximate frequency and then evaluates the result to see if the desired mode has been obtained.

Example 7.8 Estimate the fundamental period of the four-story steel frame using the Holzer method.

For years, building codes have suggested that the fundamental mode of a moment frame building be estimated as $0.1N$, where N is the number of stories. However, for a bare steel frame, the estimated period is probably closer to $0.2N$. If we use this initial approximation, the period becomes 0.8 sec, and the circular frequency $\omega = 2\pi/0.8 = 7.8$ rad/sec ≈ 8 rad/sec.

First Iteration: $f_i = \omega^2 m_i v_i$; $\omega^2 = 64$

Floor	Mass	Displacement	Inertia	Shear	Stiffness	Deflection
4	0.761	1.0	48.7			
				48.7	588	0.083
3	0.952	0.92	56.0			
				104.7	588	0.178
2	0.952	0.74	45.1			
				149.8	734	0.204
1	0.958	0.54	32.9			
				182.7	492	0.371
Base						0.165

Because the base displacement is greater than zero, the frequency needs to be increased, so another iteration is required (try $\omega = 9$).

Second Iteration: $f_i = \omega^2 m_i v_i$; $\omega^2 = 81$

Floor	Mass	Displacement	Inertia	Shear	Stiffness	Deflection
4	0.761	1.0	61.6			
				61.6	588	0.105
3	0.952	0.92	69.0			
				130.6	588	0.222
2	0.952	0.74	51.9			
				182.5	734	0.249
1	0.958	0.54	32.9			
				215.4	438	0.438
Base						−0.014

Using Equation 7.45,

$$\frac{81 - 64}{0.165 - (-0.014)} = \frac{\Delta\omega^2}{-0.014 - 0}$$

which leads to

$$\Delta\omega^2 = -0.014 \left(\frac{17}{0.179} \right) = 1.33$$

Furthermore, $\omega_3^2 = 81 - 1.33 = 79.67$ and $\omega = 8.92$ and $T = 0.704$ sec. This result can be verified by performing a third iteration with the preceding properties.

Third Iteration: $f_i = \omega^2 m_i v_i$; $\omega^2 = 79.67$

Floor	Mass	Displacement	Inertia	Shear	Stiffness	Deflection
4	0.761	1.0	60.6			
				60.6	588	0.103
3	0.952	0.897	68.0			
				128.6	588	0.219
2	0.952	0.678	51.4			
				180.0	734	0.245
1	0.958	0.433	33.0			
				213.0	438	0.433
Base						0.00

This result implies that this is the best estimate of the fundamental mode.

7.8 AXIAL LOAD EFFECTS (LINEAR APPROXIMATION)

In Section 2.6.4, it was shown that an axial load component can reduce the lateral stiffness of the member and therefore increase the tendency for lateral buckling. In this section, it is assumed that the forces that tend to cause buckling are constant and are not significantly affected by the dynamic response. When these forces vary significantly with time, they result in a time-varying stiffness property, which causes a nonlinear response, and the analysis procedure that is based on the principle of superposition is no longer valid. The simplest approximation that addresses this problem is one in which it is assumed that all axial forces are acting in an auxiliary system consisting of rigid bar segments connected by hinges.

The forces required for equilibrium in a typical segment are shown in Figure 7.6.

Taking moments about node i results in

$$N_i \left(v_j - v_i \right) = -T_j l_i$$

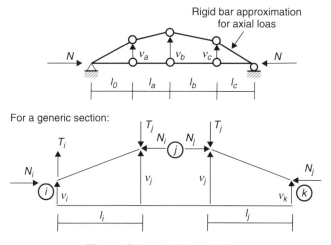

Figure 7.6   Axial load effect

This can be written as

$$-T_j = \frac{v_j - v_i}{l_i} N_i$$

In a similar manner, taking moments about node k gives

$$-T_j = \frac{v_j - v_k}{l_j} N_j$$

Combining the two expressions for T_j and rearranging some terms, we have

$$T_j = \frac{N_i}{l_i} v_i - \left(\frac{N_i}{l_i} + \frac{N_j}{l_j} \right) v_j + \frac{N_j}{l_j} v_k \qquad (7.46)$$

By direct superposition,

$$\begin{Bmatrix} T_1 \\ T_2 \\ T_3 \\ T_4 \\ \vdots \end{Bmatrix} = \begin{bmatrix} \dfrac{N_0}{l_0} + \dfrac{N_1}{l_1} & -\dfrac{N_1}{l_1} & 0 & 0 \\ -\dfrac{N_1}{l_1} & \dfrac{N_1}{l_1} + \dfrac{N_2}{l_2} & -\dfrac{N_2}{l_2} & 0 \\ 0 & -\dfrac{N_2}{l_2} & \dfrac{N_2}{l_2} + \dfrac{N_3}{l_3} & -\dfrac{N_3}{l_3} \end{bmatrix} \begin{Bmatrix} v_1 \\ v_2 \\ v_3 \\ v_4 \\ \vdots \end{Bmatrix} \qquad (7.47)$$

$$\{T\} = \begin{bmatrix} K_G \end{bmatrix} \{v\}$$

where $K_G =$ the geometric stiffness matrix

Combining this result with the elastic stiffness matrix results in the stiffness of the system, which includes the linear approximation of the effect of axial load on the system stiffness.

Example 7.9 Four-Story Steel Frame: Axial Load Effects Determine the effect of axial load on the stiffness of the four-story steel frame shown in Figure 7.7.

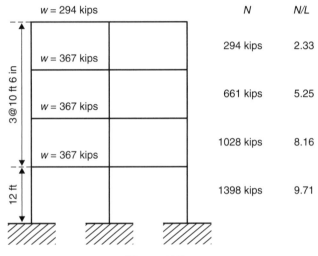

Figure 7.7

$$[K_G] = \begin{bmatrix} 2.33 & -2.33 & 0 & 0 \\ -2.33 & 7.58 & -5.25 & 0 \\ 0 & -5.25 & 13.41 & -8.16 \\ 0 & 0 & -8.16 & 17.87 \end{bmatrix}$$

$$[K] = \begin{bmatrix} 587.7 & -587.7 & 0 & 0 \\ -587.7 & 1175.3 & -587.7 & 0 \\ 0 & -587.7 & -1321.4 & -733.7 \\ 0 & 0 & -733.7 & 1225.3 \end{bmatrix}$$

$$[K] - [K_G] = \begin{bmatrix} 585.4 & -585.4 & 0 & 0 \\ -585.4 & 1167.7 & -582.5 & 0 \\ 0 & 582.5 & 1308.0 & -725.6 \\ 0 & 0 & -725.6 & 1207.4 \end{bmatrix}$$

7.9   MODAL EQUATIONS FOR UNDAMPED TIME-DEPENDENT FORCE ANALYSIS

The equations of motion for the forced-vibration response of an undamped MDOF system can be written in matrix form as

$$[m]\{\ddot{v}\} + [k]\{v\} = \{p(t)\} \tag{7.48}$$

The geometric coordinates can be obtained from the normal (generalized) coordinates using the modal matrix:

$$\{v\} = [\phi]\{Y\} \qquad \{\ddot{v}\} = [\phi]\{\ddot{Y}\} \tag{7.49}$$

Substitution of this coordinate transformation into the equations for dynamic equilibrium given previously results in

$$[m][\phi]\{\ddot{Y}\} + [k][\phi]\{Y\} = \{p(t)\} \tag{7.50}$$

Now premultiply the above equation by $\{\phi_n\}^T$ to obtain

$$\{\phi_n\}^T [m][\phi]\{\ddot{Y}\} + \{\phi_n\}^T [k][\phi]\{Y\} = \{\phi_n\}^T \{p(t)\} \tag{7.51}$$

Using the orthogonality properties of normal modes, we can rewrite the preceding equation as

$$\{\phi_n\}^T [m]\{\phi_n\}\ddot{Y}_n + \{\phi_n\}^T [k]\{\phi_n\}Y_n = \{\phi_n\}^T \{p(t)\} \tag{7.52}$$

because all the terms except the nth will be zero. This modal equation for the nth mode can also be written as

$$M_n^* \ddot{Y}_n + K_n^* Y_n = P_n^*(t) \tag{7.53}$$

where $M_n^* = \{\phi_n\}^T [m]\{\phi_n\} = \sum m_i \phi_{in}^2 = $ the generalized mass in the nth mode

$K_n^* = \left\{\phi_n^T\right\} [k]\phi_n = \sum k_i \phi_{in}^2 = $ the generalized stiffness for the nth mode

$P_n^*(t) = \{\phi_n\}^T \{p(t)\} = \sum p_i \phi_{in} = $ the generalized applied force for the nth mode

The preceding modal equation can be further simplified by dividing through by M_n^* for the nth mode of vibration to obtain

$$\ddot{Y}_n + \omega_n^2 Y_n = \frac{P_n^*(t)}{M_n^*} \tag{7.54}$$

where

$$\omega_n^2 = \frac{K_n^*}{M_n^*}$$

Because both K_n^* and M_n^* are diagonal matrices, due to the orthogonality property of normal modes, the matrix product $[M^*]^{-1}[K^*]$ will be a diagonal matrix of ω^2.

Note that the previous modal equations of motion for the forced vibration of an undamped system are completely uncoupled and can be solved separately. Also, many time-dependent forces are for short-duration loads, and hence the damping can be neglected. Each modal equation is identical in form to the equation of motion for forced vibration of an undamped linear SDOF system defined previously. Therefore, any of the numerical methods developed previously are applicable. This process has been summarized by Clough and Penzien[1] as follows: "The use of normal coordinates serves to transform the equations of motion from a set of N simultaneous differential equations that are coupled by off-diagonal terms in the mass and stiffness matrices, to a set of N independent normal coordinate equations." The complete solution for the system is then obtained by superimposing the independent modal solutions. For this reason, the method is often referred to as the *modal superposition method*. Use of this method also leads to significant savings in computational effort, because, in most cases, it will not be necessary to use all N modal responses to accurately represent the response of the structure. For most structural systems, the lower modes make the primary contribution to the total response. Therefore, the response can usually be represented to sufficient accuracy in terms of a limited number of modal responses in the lower modes.

Example 7.10 Time-Dependent Force Analysis Consider the four-story steel frame of the previous examples and determine the elastic dynamic displacement response of the frame to a time-dependent force applied at the roof level having a zero rise time and a constant amplitude,

[1]R. Clough and J. Penzien, *Dynamics of Structures* (New York: McGraw-Hill, 1975).

as shown in Figure 7.8. The steps required are as follows. The first three steps were completed in previous examples, as indicated in parentheses:

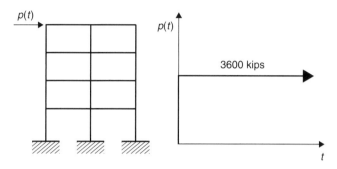

Figure 7.8

a. Solve for the frequencies (Example 7.1).
b. Determine the shapes of the four modes of vibration (Example 7.2).
c. Calculate the generalized mass for the four modes (Example 7.4).
d. Calculate the generalized forces.

$$\{P^*\} = [\phi]^T\{P(t)\}$$

$$= \begin{bmatrix} 1.00 & 0.90 & 0.68 & 0.43 \\ 1.00 & 0.19 & -0.82 & -0.96 \\ 1.00 & -1.05 & -0.39 & 0.94 \\ 1.00 & -2.25 & 3.14 & -2.20 \end{bmatrix} \begin{Bmatrix} 3600 \\ 0 \\ 0 \\ 0 \end{Bmatrix} = \begin{Bmatrix} 3600 \\ 3600 \\ 3600 \\ 3600 \end{Bmatrix}$$

e. Set up the equations of motion:

$$M_n^*\ddot{Y}_n + K_n^*Y_n = P_n^*$$

which can be written as

$$\ddot{Y}_n + \omega_n^2 Y_n = \frac{P_n^*}{M_n^*}$$

With ω_n^2 and M_n^* determined previously, the generalized stiffness can be determined as the product $K_n^* = \omega_n^2 M_n^*$. For

$$K_1^* = \omega_1^2 M_1^* = (8.94)^2 * 2.15 = 172 \text{ kips/in}$$
$$K_2^* = \omega_2^2 M_2^* = (25)^2 * 2.32 = 1453 \text{ kips/in}$$

$$K_3^* = \omega_3^2 M_3 = (39.9)^2 * 2.80 = 4464 \text{ kips/in}$$

$$K_4^* = \omega_4^2 M_4 = (48.68)^2 * 19.58 = 46,400 \text{ kips/in}$$

$$Y_{np} = \frac{P_n^*}{K_n^*} = \frac{P_n^*}{\omega_n^2 M_n^*} = Y_n(\text{static}) = \left\{ \begin{array}{c} \dfrac{3600}{80*2.14} \\ \dfrac{3600}{626*2.30} \\ \dfrac{3600}{1594*2.80} \\ \dfrac{3600}{2370*19.6} \end{array} \right\} = \left\{ \begin{array}{c} 21.0 \\ 2.69 \\ 0.81 \\ 0.08 \end{array} \right\}$$

f. Determine the solution of the equations of motion assuming starting-at-rest initial conditions:

$$y = A \sin \omega t + B \cos \omega t + y_p \qquad y(0) = 0 \Rightarrow B = -y_p$$

$$\dot{y} = A\omega \cos \omega t - B\omega \sin \omega t \qquad \dot{y}(0) = 0 \Rightarrow A = 0$$

$$y = -y_p \cos \omega t + y_p = y_p(1 - \cos \omega t)$$

General Solution:

$$Y_n(t) = Y_{np}\left(1 - \cos \omega_n t\right)$$

$$Y_n(t) = \left\{ \begin{array}{c} 21.0\,(1 - \cos \omega_1 t) \\ 2.69\,(1 - \cos \omega_2 t) \\ 0.81\left(1 - \cos \omega_3 t\right) \\ 0.08\,(1 - \cos \omega_4 t) \end{array} \right\}$$

g. Transform to geometric coordinates:

$$\{v_n\} = [\phi]\{Y\}$$

$$= \begin{bmatrix} 1.00 & 1.00 & 1.00 & 1.00 \\ 0.90 & 0.19 & -1.05 & -2.25 \\ 0.68 & -0.82 & -0.39 & 3.14 \\ 0.43 & -0.96 & 0.94 & -2.20 \end{bmatrix} \left\{ \begin{array}{c} 21.0\,(1 - \cos \omega_1 t) \\ 2.69\,(1 - \cos \omega_2 t) \\ 0.81\left(1 - \cos \omega_3 t\right) \\ 0.08\,(1 - \cos \omega_4 t) \end{array} \right\}$$

$\{v(t)\} =$

$$\begin{Bmatrix} 21.0(1 - \cos\omega_1 t) + 2.69(1 - \cos\omega_2 t) + 0.81(1 - \cos\omega_3 t) + 0.08(1 - \cos\omega_4 t) \\ 18.9(1 - \cos\omega_1 t) + 0.51(1 - \cos\omega_2 t) - 0.95(1 - \cos\omega_3 t) - 0.18(1 - \cos\omega_4 t) \\ 14.3(1 - \cos\omega_1 t) - 2.21(1 - \cos\omega_2 t) - 0.32(1 - \cos\omega_3 t) + 0.25(1 - \cos\omega_4 t) \\ 9.03(1 - \cos\omega_1 t) - 2.58(1 - \cos\omega_2 t) + 0.76(1 - \cos\omega_3 t) - 0.18(1 - \cos\omega_4 t) \end{Bmatrix}$$

Example 7.11(M)   Use MATLAB's Symbolic Math Toolbox to solve Example 7.10, given the following general solution provided in that example:

$$Y_n(t) = Y_{np}\left(1 - \cos\omega_n t\right)$$

Calculate the generalized forces using what we know from previous examples:

```
>> M=[0.761 0 0 0; 0 0.952 0 0; 0 0 0.952 0; 0 0 0 0.958];
>> K=500.*[1.175 -1.175 0 0; -1.175 2.351 -1.175 0;
     0 -1.175 2.643 -1.467; 0 0 -1.467 2.451];
>> [Eigenvectors, Eigenvalues] = eig(K,M);
>> phi= [1.0000    1.0000    1.0000    1.0000;
         0.8965    0.1882   -1.0609   -2.0784;
         0.6777   -0.8145   -0.3876    2.8453;
         0.4327   -0.9557    0.9518   -1.9853];
>> Pt= [3600; 0; 0; 0]
Pt =
        3600
           0
           0
           0
>> Pstar=phi'*Pt
Pstar =
        3600
        3600
        3600
        3600
>> Mstar =phi'*M*phi
Mstar =
     2.1427   -0.0000    0.0000   -0.0001
    -0.0000    2.3013    0.0000    0.0000
     0.0000    0.0000    2.8434   -0.0000
    -0.0001    0.0000   -0.0000   16.3564
>> Kstar = Eigenvalues'*Mstar
Kstar =
    1.0e+04  *
     0.0171   -0.0000    0.0000   -0.0000
    -0.0000    0.1442    0.0000    0.0000
     0.0000    0.0000    0.4524   -0.0000
    -0.0000    0.0000   -0.0000    3.8872
```

```
>> YNP = inv(Kstar)*Pstar
YNP =
   21.0286
    2.4965
    0.7955
    0.0927
```

Define the needed symbolic parameters (you may not need to redefine some of these variables if you are continuing the script from previous examples):

```
>> syms t w1 w2 w3 w4 Yt Ynt
>> Yt =[1-cos(w1*t); 1-cos(w2*t); 1-cos(w3*t); 1-cos(w4*t)]

Yt =

 1 - cos(t*w1)
 1 - cos(t*w2)
 1 - cos(t*w3)
 1 - cos(t*w4)
>>  Ynt=vpa(YNP.*Yt,4)

Ynt =

     21.03 - 21.03*cos(t*w1)
     2.496 - 2.496*cos(t*w2)
    0.7955 - 0.7955*cos(t*w3)
  0.09274 - 0.09274*cos(t*w4)
>> vt =vpa(phi*Ynt,4)

vt =

 24.41 - 2.496*cos(t*w2) - 0.7955*cos(t*w3) - 0.09274*cos(t*w4)
       - 21.03*cos(t*w1)
 0.8439*cos(t*w3) - 0.4698*cos(t*w2) - 18.85*cos(t*w1)
       + 0.1928*cos(t*w4) + 18.29
 2.033*cos(t*w2) - 14.25*cos(t*w1) + 0.3083*cos(t*w3)
       - 0.2639*cos(t*w4) + 12.17
 2.386*cos(t*w2) - 9.099*cos(t*w1) - 0.7571*cos(t*w3)
       + 0.1841*cos(t*w4) + 7.286
```

Notice that the result, if reordered and factored, is basically the same as that in Example 7.10. Small numerical differences are due to round-off errors in hand calculations.

Example 7.12 Modal Analysis of a Building Subjected to Air Blast The reflected pressure on the side of a building subjected to an air blast (explosion) at the ground surface can be represented as having a triangular distribution with a maximum pressure at the base, which then diminishes to near zero as it propagates up the face of the building, as shown in Figure 7.9. The pressure load is then transformed into concentrated loads at the floor levels of the frame. The duration of the loading, t_1, is taken as 0.35 sec. Because of the short duration of the loading, damping can be neglected, and the maximum response can be estimated as follows:

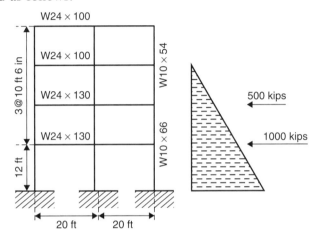

Figure 7.9

$$Y_n(t) = D_n \frac{P_{0n}}{K_n} \sin \omega_n t \qquad (7.55)$$

where D_n is a dynamic load factor based on the load duration relative to the modal period, as shown in Figure 7.10a and 7.10a:

$$P_{0n} = \begin{Bmatrix} P_{01}(t) \\ P_{02}(t) \\ P_{03}(t) \\ P_{04}(t) \end{Bmatrix} = \{\phi_n\}^T \begin{Bmatrix} 0 \\ 0 \\ 1 \\ 2 \end{Bmatrix} 500 = \begin{Bmatrix} 770 \\ -1370 \\ 745 \\ -630 \end{Bmatrix} \text{ kips}$$

$$\begin{Bmatrix} \dfrac{t_1}{T_1} \\ \dfrac{t_1}{T_2} \\ \dfrac{t_1}{T_3} \\ \dfrac{t_1}{T_4} \end{Bmatrix} = \frac{0.35}{2\pi} \begin{Bmatrix} \omega_1 \\ \omega_2 \\ \omega_3 \\ \omega_4 \end{Bmatrix} = \begin{Bmatrix} 0.50 \\ 1.40 \\ 2.23 \\ 2.73 \end{Bmatrix}$$

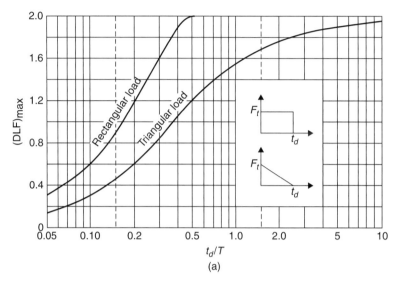

Figure 7.10a Maximum dynamic load factor, D for an undamped, elastic SDOF

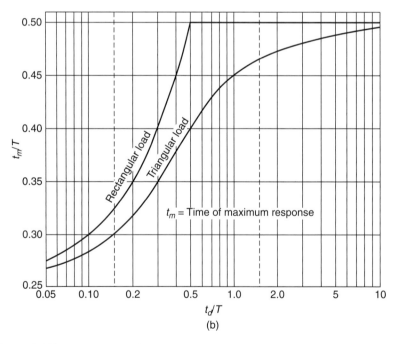

Figure 7.10b Time to maximum response, t_m, for an undamped elastic SDOF[2]

[2]US Army Corps of Engineers, *Design of Structures to Resist the Effects of Atomic Weapons*, Manual EM 1110-345-415, 1957.

$$\begin{Bmatrix} D_1 \\ D_2 \\ D_3 \\ D_4 \end{Bmatrix} = \begin{Bmatrix} 1.2 \\ 1.68 \\ 1.76 \\ 1.80 \end{Bmatrix}$$

$$\omega_n = \begin{Bmatrix} 8.9 \\ 25.0 \\ 39.9 \\ 48.7 \end{Bmatrix} \quad M_n^* = \begin{Bmatrix} 2.15 \\ 2.32 \\ 2.80 \\ 19.6 \end{Bmatrix} \quad K_n^* = \omega_n^2 M_n^* = \begin{Bmatrix} 172 \\ 1450 \\ 4458 \\ 46438 \end{Bmatrix}$$

Substituting these results into Equation 7.55 for global displacement yields the following:

$$Y_1(t) = 1.2 * 770/172 = 5.44 \sin 8.9t$$
$$Y_2(t) = 1.68 * (-1370)/1450 = -1.59 \sin 25t$$
$$Y_3(t) = 1.76 * 745/4458 = 0.29 \sin 39.9t$$
$$Y_4(t) = 1.80 * (-630)/46438 = -0.024 \sin 48.7t$$

The relative displacements can be calculated as

$$\{v(t)\} = [\phi]\{Y(t)\}$$

$$\{v(t)\} = \begin{bmatrix} 1.00 & 1.00 & 1.00 & 1.00 \\ 0.90 & 0.19 & -1.05 & -2.25 \\ 0.68 & -0.82 & -0.39 & 3.14 \\ 0.43 & -0.96 & 0.94 & -2.20 \end{bmatrix} \begin{Bmatrix} 5.44 \ \sin 8.9t \\ -1.59 \ \sin 25t \\ 0.29 \ \sin 39.9t \\ -0.024 \ \sin 48.7t \end{Bmatrix}$$

In a similar manner, the elastic forces developed in the structure under the blast loading can be evaluated as

$$\{f\} = [\phi]\{m\ddot{Y}\} = [\phi]\{m\omega^2 Y\}$$

$$\{f(t)\} = \begin{bmatrix} 1.00 & 1.00 & 1.00 & 1.00 \\ 0.90 & 0.19 & -1.05 & -2.25 \\ 0.68 & -0.82 & -0.39 & 3.14 \\ 0.43 & -0.96 & 0.94 & -2.20 \end{bmatrix} \begin{Bmatrix} 2.15 * 79.2 * 5.44 \sin 8.9t \\ -625 * 2.32 * 1.59 \sin 25t \\ 2.8 * 1592 * 0.29 \sin 39.9t \\ -19.6 * 2372 * 0.24 \sin 48.7t \end{Bmatrix}$$

Example 7.13(M) Solve Example 7.12 using MATLAB's Symbolic Math Toolbox.

First, we define MATLAB variables representing the entities defined in Example 7.12. We assume that the vectors and matrices generated in the previous example are still available on your MATLAB workspace and then we apply the formulas:

```
>> syms Ysin, Ynt2
>> Ysin = [sin(w1*t); sin(w2*t); sin(w3*t); sin(w4*t)]

Ysin =

 sin(t*w1)
 sin(t*w2)
 sin(t*w3)
 sin(t*w4)

>> Ynt2 =vpa(inv(Kstar)*Dn.*P0n.*Ysin, 4)

Ynt2 =

    5.408*sin(t*w1)
   -1.588*sin(t*w2)
    0.2948*sin(t*w3)
  -0.02608*sin(t*w4)

>> vt =vpa(phi*Ynt2, 4)

vt =

 5.408*sin(t*w1) - 1.588*sin(t*w2) + 0.2948*sin(t*w3)
                 - 0.02608*sin(t*w4)
 4.848*sin(t*w1) - 0.2988*sin(t*w2) - 0.3128*sin(t*w3)
                 + 0.0542*sin(t*w4)
  3.665*sin(t*w1) + 1.293*sin(t*w2) - 0.1143*sin(t*w3)
                 - 0.0742*sin(t*w4)
  2.34*sin(t*w1) + 1.517*sin(t*w2) + 0.2806*sin(t*w3)
                 + 0.05177*sin(t*w4)
```

7.10 MODAL EQUATIONS OF DAMPED FORCED VIBRATION

If viscous damping is included, the equations of motion become

$$[M]\{\ddot{v}\} + [C]\{\dot{v}\} + [K]\{v\} = \{P(t)\} \qquad (7.56)$$

When these are transformed into modal coordinates, the mass matrix, $[M^*]$, and the stiffness matrix, $[K^*]$, are diagonal matrices, and the inertia forces and the elastic restoring forces are uncoupled. However, applying the same transformation to the damping matrix, $[C^*]$, results in terms that are coupled in the velocity term. One possibility for getting around this problem is to assume the off-diagonal terms are zero, whether they are or not. However, the off-diagonal terms are generally small relative to terms on the main diagonal. Another possibility is to assume mass proportional damping, which is easy to use but results in a lower fraction of critical damping in the higher modes than for the fundamental mode. As mentioned previously for the SDOF system, this is counter to observed behavior. Another alternative might be the use of stiffness proportional damping in which the higher modes decay quicker than the fundamental mode. For an elastic system, it is possible to determine the modes for the undamped system and then insert a fraction of critical damping into each modal equation.

It would be helpful if conditions could be established in which uncoupling of these terms in the damping matrix would occur. In 1945, Rayleigh showed that a damping matrix that is a linear combination of the mass and stiffness matrices will satisfy the orthogonality condition:

$$[C] = \alpha_0[M] + \alpha_1[K] \qquad (7.57)$$

where α_0 and α_1 are arbitrary proportionality factors. Consider a damping matrix of that form and premultiply by $\{\phi_n\}^T$ and postmultiply by $[\phi]$ to obtain

$$\{\phi\}^T[C][\phi] = \alpha_0\left\{\phi_n^T\right\}[M][\phi] + \alpha_1\left\{\phi_n^T\right\}[K][\phi] \qquad (7.58)$$

Using the orthogonality conditions, we obtain

$$C_n = \{\phi_n\}^T[C]\{\phi_n\} = \alpha_0\{\phi_n\}^T[M]\{\phi_n\} + \alpha_1\{\phi_n\}^T[K]\{\phi_n\} \qquad (7.59)$$

which can also be written as

$$C_n = \left(\alpha_0 + \alpha_1\omega_n^2\right) M_n \qquad (7.60)$$

which is uncoupled. In general, Rayleigh damping can be expressed as

$$[C] = [M]\sum_i a_i \left([M]^{-1}[K]\right)^i \qquad (7.61)$$

where $i =$ the number of terms in the series or the number of
conditions to be satisfied

By taking only a single term of the Rayleigh damping expression, it was shown previously that

$$c = \alpha_0 m = 2\xi m\omega$$

which implies

$$\alpha_0 = 2\xi\omega$$

In this representation, there is only one condition to satisfy and therefore

$$\xi = \frac{\alpha_0}{2\omega} = \frac{\alpha_0 T}{4\pi} \tag{7.62}$$

This result indicates that the damping coefficient increases linearly with the period. While this representation is easy to accommodate from a computational sense, it has been noted earlier that it leads to a reduced percentage of damping in the higher modes, which is known to be inconsistent with the measured response.

An alternative that also avoids the problem of velocity coupling is to use the second term of the Rayleigh expression, which has the form

$$c = \alpha_1 \omega^2 m = 2\xi m\omega$$

which implies

$$\alpha_1 = \frac{2\xi}{\omega}$$

As with mass proportional damping, there is only one condition to satisfy and

$$\xi = \frac{\alpha_1 \omega}{2} = \frac{\alpha_1 \pi}{T} \tag{7.63}$$

This is known as *stiffness proportional damping*, and it is more intuitive because the higher modes tend to decay more rapidly than the fundamental mode. Rayleigh damping considers two terms of the general expansion and hence has two conditions to satisfy, which are the damping values in two modes of vibration. By taking into account the first two terms, the general expression for damping can be simplified to

$$c_n = \sum_i a_i \omega_n^{2i} m_n = 2\xi_n \omega_n m_n \tag{7.64}$$

and

$$\xi_n = \frac{1}{2\omega_n} \sum_i a_i \omega_n^{2i} = \frac{1}{2}[Q]\{a\} \qquad (7.65)$$

where

$$\{a\} = 2[Q]^{-1}\{\xi\} \quad \text{and} \quad Q_i = \frac{\omega_n^{2i}}{\omega_n}$$

$$[Q] = \begin{bmatrix} \dfrac{1}{\omega_1} & \omega_1 \\[2ex] \dfrac{1}{\omega_2} & \omega_2 \end{bmatrix}$$

$$[Q]^{-1} = \begin{bmatrix} \omega_2 & -\omega_1 \\[2ex] -\dfrac{1}{\omega_2} & \dfrac{1}{\omega_1} \end{bmatrix} \dfrac{\omega_1 \omega_2}{\omega_2^2 - \omega_1^2}$$

This results in an expression for the two constants a_0 and a_1 in terms of the two circular frequencies, where the percentage of critical damping is specified, and their respective damping ratios

$$\begin{Bmatrix} a_0 \\ a_1 \end{Bmatrix} = 2 \begin{bmatrix} \omega_2 & -\omega_1 \\[2ex] -\dfrac{1}{\omega_2} & \dfrac{1}{\omega_1} \end{bmatrix} \dfrac{\omega_1 \omega_2}{\omega_2^2 - \omega_1^2} \begin{Bmatrix} \xi_1 \\ \xi_2 \end{Bmatrix} \qquad (7.66)$$

Once the constants a_0 and a_1 are determined, the resulting damping in other modes can be determined from

$$\xi_k = \frac{\alpha_0}{2\omega_k} + \frac{\alpha_1 \omega_k}{2} \qquad (7.67)$$

With the addition of more terms to the general expression, it is possible to compute the damping coefficients necessary to provide a decoupled system having any desired damping ratio in any specified number of modes. However, this may not be particularly meaningful in view of the inherent uncertainty in the damping matrix. A convenient representation of damping in elastic MDOF systems is to find the modes for the undamped system and insert a fraction of critical damping into each modal equation.

Example 7.14   For the four-story steel frame, determine the damping matrix so as to give 5 percent of critical damping in the first and third modes. What is the damping that is induced into the second and fourth modes with this representation of damping? Plot the resulting damping function.

From the earlier solution for the mode shapes and frequencies:

$$\omega_1 = 8.94 \text{ rad/sec} \qquad \omega_2 = 25.03 \text{ rad/sec}$$

$$\omega_3 = 39.93 \text{ rad/sec} \qquad \omega_4 = 48.68 \text{ rad/sec}$$

$$\left\{ \begin{array}{c} \alpha_0 \\ \alpha_1 \end{array} \right\} = 2 \left[\begin{array}{cc} 39.93 & -8.94 \\ \dfrac{1}{39.93} & \dfrac{1}{8.94} \end{array} \right] \dfrac{8.94 * 39.93}{39.93^2 - 8.94^2} \left\{ \begin{array}{c} 0.05 \\ 0.05 \end{array} \right\} = \left\{ \begin{array}{c} 0.731 \\ 0.002 \end{array} \right\}$$

$$[C] = 0.731[M] + 0.002[K]$$

ω	T	ξ
8.94	0.702	0.0495
25.03	0.251	0.0400
39.93	0.157	0.0491
48.68	0.129	0.0560

The results of using these equations produce the values for the damping function given in the preceding table and shown in Figure 7.11.

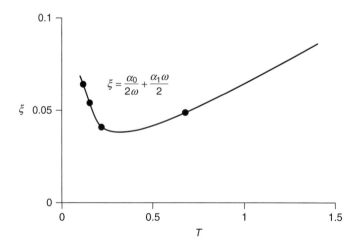

Figure 7.11   Damping function for a four-story steel building

7.11   MODAL EQUATIONS FOR SEISMIC RESPONSE ANALYSIS

For earthquake analysis, the time-dependent force must be replaced with the effective time-dependent force, $P_e(t)$. The displacement of an MDOF structure relative to the base is v, and the ground displacement is v_g. The total displacement of the structure can then be represented as the sum of the displacement relative to the ground plus the ground displacement:

$$\{u\} = \text{total displacement} = \{v\} + \{1\}v_g$$

The differential equation of motion for response to ground motion becomes

$$[M]\left\{\{\ddot{v}\} + \{1\}\ddot{v}_g\right\} + [C]\{\dot{v}\} + [K]\{v\} = 0 \qquad (7.68)$$

These equations can also be written in the form

$$[M]\{\ddot{v}\} + [C]\{\dot{v}\} + [K]\{v\} = -[M]\{1\}\ddot{v}_g(t) = P_e(t) \qquad (7.69)$$

where the effective dynamic force is given by the product of the mass at any level and the base acceleration:

$$P_e(t) = [M]\{\Gamma\}\ddot{v}_g(t) \qquad (7.70)$$

Here $\{\Gamma\}$ is a vector of influence coefficients of which component i represents the acceleration at displacement coordinate i caused by a unit ground acceleration at the base. For the structural model in which the degrees of freedom are represented by the horizontal displacements of the story levels, the vector $\{\Gamma\}$ becomes a unity vector, [1], because for a unit ground acceleration in the horizontal direction all degrees of freedom have a unit horizontal acceleration. As done previously for the time-dependent forcing function, the geometric coordinates can be obtained from the normal (generalized) coordinates using the modal transformation:

$$\{v\} = [\phi]\{Y\} \quad \{\dot{v}\} = [\phi]\{\dot{Y}\} \quad \{\ddot{v}\} = [\phi]\{\ddot{Y}\} \qquad (7.71)$$

Substitution of this coordinate transformation into the equations of dynamic equilibrium results in

$$[M][\phi]\left\{\ddot{Y}\right\} + [C][\phi]\left\{\dot{Y}\right\} + [K][\phi]\{Y\} = \left\{P_e(t)\right\} \qquad (7.72)$$

Now premultiply the preceding equation by $\{\phi_n\}^T$ and use the orthogonality property of normal mode shapes to obtain

$$\{\phi_n\}^T [M]\{\phi_n\}\ddot{Y}_n + \{\phi_n\}^T [C]\{\phi_n\}\dot{Y}_n$$
$$+ \{\phi_n\}^T [K]\{\phi_n\} \{Y\} = \{\phi_n\}^T \{P_e(t)\} \tag{7.73}$$

Because all the terms except the nth term will be zero, this modal equation for the nth mode can also be written as

$$M_n^* \ddot{Y}_n + C_n^* \dot{Y}_n + K_n^* Y_n = P_n^*(t) \tag{7.74}$$

or, alternatively, as

$$\ddot{Y}_n + 2\xi_n \omega_n \dot{Y}_n + \omega_n^2 Y_n = -\frac{P_{en}^*}{M_n^*}\ddot{v}_g$$

where the ratio $\frac{P_{en}^*}{M_{n*}}$ is often referred to as the *modal participation ratio*.

Example 7.15 Modal Response to Earthquake Input   Develop the equations of motion for the four-story steel frame subjected to earthquake ground motion, neglecting damping. Consider the alternate form of the equations of motion with the damping omitted:

$$\Gamma_p = \frac{\{\phi_p\}^T [M][1]}{M_p^*} \qquad p = 1, 2, 3, 4$$

$$\Gamma_1 = \frac{2.68}{2.15} = 1.25 \quad \Gamma_2 = \frac{-0.76}{2.32} = -0.33$$
$$\Gamma_3 = \frac{0.29}{2.80} = 0.10 \quad \Gamma_4 = \frac{-0.50}{19.6} = -0.06$$

The responses for each mode to the earthquake input become

$$\ddot{Y}_1 + 79.9Y_1 = -1.25\ddot{v}_g(t)$$
$$\ddot{Y}_2 + 626Y_2 = 0.33\ddot{v}_g(t)$$
$$\ddot{Y}_3 + 1594Y_2 = -0.10\ddot{v}_g(t)$$
$$\ddot{Y}_4 + 2370Y_4 = 0.06\ddot{v}_g(t)$$

PROBLEMS

Problem 7.1

Determine the natural frequencies and mode shapes of the system shown in Figure 7.12.

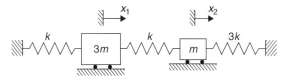

Figure 7.12

Problem 7.2

The rigid bar shown in Figure 7.13 is supported by two springs, one on each end. Using the displacement coordinates at the center of the bar (a vertical displacement and a rotation), write the equations of motion for free vibration in matrix form.

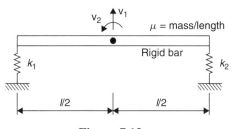

Figure 7.13

Problem 7.3

One moment-resisting frame from a three-story building structure is shown in Figure 7.14. Consider this as a three-degree-of-freedom system with a lateral degree of freedom at each story level. Assume its mass is concentrated at the floor levels and that the girders are rigid relative to the columns. Neglect damping and the effects of gravity forces on column bending moments and column stiffness. Develop the inertia and stiffness matrices, $[M]$ and $[K]$. Also find the frequency equation and the corresponding three frequencies. Determine and sketch the three mode shapes. Use $E = 29,000$ ksi.

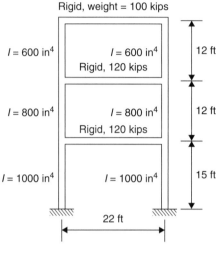

Rigid, weight = 100 kips

$I = 600$ in^4 $I = 600$ in^4 12 ft
Rigid, 120 kips

$I = 800$ in^4 $I = 800$ in^4 12 ft
Rigid, 120 kips

$I = 1000$ in^4 $I = 1000$ in^4 15 ft

22 ft

Figure 7.14

Problem 7.4(M)

Use MATLAB to calculate the natural frequencies and mode shapes of the frame of Problem 7.3. Can you use MATLAB to plot the mode shapes and compare them to the results you obtained in Problem 7.3?

Problem 7.5

Consider the six-story steel office building of Problem 3.1. Determine the change in the fundamental period if second-order P-Δ effects are estimated by including the linear geometric (string) stiffness.

Problem 7.6(M)

Use MATLAB to solve Problem 7.5.

Problem 7.7

Consider the six-story steel office building of Problem 3.1. Determine the mode shapes and frequency for the second mode of vibration using the Holzer method. Assume that the period of the second mode is approximately 0.5 sec and proceed to refine your estimate using the Holzer method with 1 in for the displacement amplitude at the roof.

CHAPTER 8

NONLINEAR RESPONSE OF MULTIPLE-DEGREE-OF-FREEDOM SYSTEMS

In Chapter 7, we assumed that the structures were linear and that therefore the response could be represented by linear influence coefficients. This made it possible to compute the vibration mode shapes and frequencies for the structure and to evaluate the elastic response in terms of the modal coordinates. However, any change in the physical constants due to nonlinear behavior will alter the vibration characteristics of the system, and the normal coordinate uncoupling of the equations of motion will not be possible.

In the analysis of structural systems for dynamic loads, there are many instances in which the physical properties of the system can no longer be assumed to remain constant during the dynamic response. Any moderate to strong earthquake ground motion will drive a structure designed by conventional methods into the inelastic range, particularly in certain critical regions of the structure. The most common source of nonlinear behavior occurs when the elastic limit of the material is exceeded. However, changes in the axial forces in the structural members, such as columns combined with increased lateral displacements, can cause changes in the geometric stiffness coefficients. Changes in the lateral stiffness coefficients can also occur in structures that are isolated from the base or some other part of the structure and in aboveground pipelines that may be exposed to significant temperature changes in addition to ground motions.

8.1 STATIC NONLINEAR ANALYSIS

Often it is necessary to estimate the capacity of a structure to resist a dynamic load that will force the structure beyond the elastic limit. In order to make this estimate, a static nonlinear (pushover) analysis is often conducted. The representative ultimate load is divided into a number of small load increments that are applied sequentially to the structure. As nonlinear behavior occurs, the incremental stiffness is modified for the next load increment. Therefore, the behavior of the nonlinear system is approximated by the response of a sequential series of linear systems having varying stiffness. Nonlinear static analyses can be considered as a subset of the step-by-step dynamic analyses and can use the same solution procedure without the time-related inertia forces and damping forces. If the total applied force is divided into a number of equal load increments, the equations of static equilibrium can be written in matrix form for a small load increment, i, during which the behavior of the structure is assumed to be linear elastic:

$$[K_i]\{\Delta v_i\} = \{\Delta F_i\} \tag{8.1}$$

For computational purposes, it is convenient to rewrite this equation in the following form:

$$[K_i]\{\Delta v_i\} = \{F_i\} - \{R_{i-1}\} \tag{8.2}$$

where K_t = the tangent stiffness matrix for the current load increment
$\quad F_i$ = the load at the end of the current load increment
$\quad R_{i-1}$ = the restoring force at the beginning of the load increment, which is defined as

$$\{R_{i-1}\} = \sum_{i=0}^{i-1} K_i \{\Delta v_i\} \tag{8.3}$$

For buildings the lateral force distribution is generally based on the distribution of the static equivalent lateral forces specified in building codes, which tend to approximate the first and second modes of vibration. These forces are increased in a proportional manner by a selected load factor. The lateral loading is increased until either the structure becomes unstable or a specified limit condition is attained. This type of nonlinear analysis is generally referred to as a *pushover analysis* and results in a plot of the maximum (roof) displacement versus the base shear, as shown for a six-story frame in Figure 8.1. This type of analysis can also be used to determine the sequence of hinging, which can indicate possible problem areas in the structural system. Note that the pushover curve shown in Figure 8.1 ignores the effect of geometric stiffness.

Figure 8.2[1] shows a similar analysis for the Shanghai World Financial Center, which includes the effect of geometric stiffness.

Important structural components of this building are shown in Figure 8.3, and the results of static nonlinear analysis are presented in Figure 8.4.

8.2 DYNAMIC NONLINEAR ANALYSIS

In Chapter 6, we introduced the step-by-step integration procedure for calculating the dynamic response of single-degree-of-freedom (SDOF) systems. The response time history was divided into short, equal time increments (time-steps), and the response was calculated during each increment for a linear system having the properties determined at the beginning of the time interval. This same procedure can be used for a multiple-degree-of-freedom (MDOF) system with the use of vectors and matrices for the MDOF parameters.

The equation of motion for an MDOF system without damping and subjected to a time dependent force can be expressed in matrix form as:

$$[M]\left\{\ddot{v}_{t+\Delta t}\right\} + [K]\left\{v_{t+\Delta t}\right\} = \left\{P_{t+\Delta t}\right\} \qquad (8.4)$$

If damping is included, it is convenient to include it as *Rayleigh damping,* which was discussed in Section 7.10. With this idealization, the damping relation becomes a function defined in terms of two variables, as given in Equation 7.57. Making this substitution, the equations of motion at the end of a time-step become:

$$[M]\{\ddot{v}_{t+\Delta t}\} + \alpha[M]\{\dot{v}_{t+\Delta t}\} + \beta\left[K_{in}\right]\{\dot{v}_{t+\Delta t}\}$$
$$+ \left[K_{t+\Delta t}\right]\{\Delta v\} = \left\{P_{t+\Delta t}\right\} - \left\{R_t\right\} \qquad (8.5)$$

[1]L. E. Robertson and S. T. See, "The Shanghai World Financial Center," *STRUCTURE* Magazine, June 2007.

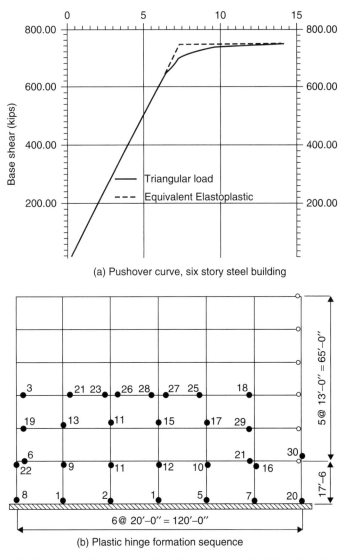

(a) Pushover curve, six story steel building

(b) Plastic hinge formation sequence

Figure 8.1   Static nonlinear analysis of a six-story building (F. Naeim, *The Seismic Design Handbook*, 2nd ed. (Dordrecht, Netherlands: Springer, 2001), reproduced with kind permission from Springer Science+Business Media B.V.)

where $\{R_t\} = \sum\limits_{\tau=0}^{t} K_\tau \Delta v_\tau$ is the resistance at the beginning of the time interval, and the damping term is based on the initial stiffness, K_{in}.

Figure 8.2   View of the building (Courtesy of Kohn Pedersen Fox)

Note that equilibrium is written at the end of the time-step. Introducing the acceleration and velocity at the end of the time-step as given by Equations 6.9 and 6.12 but in vector form, collecting terms and rearranging results in:

$$[\tilde{K}]\{\Delta v\} = \{\tilde{P}\} \tag{8.6}$$

where $\tilde{K}$ and $\tilde{P}$ denote incremental stiffness and load, respectively.

This is the pseudostatic form of the equations of motion for a multiple degree-of-freedom system. It results in a group of simultaneous equations that must be solved at the end of each time step. The time steps must be short enough to accurately represent the time dependent loading and to insure the stability of the numerical integration procedure. When used for linear systems, the procedure can be significantly simplified, since it is not necessary to modify the structural properties that remain constant

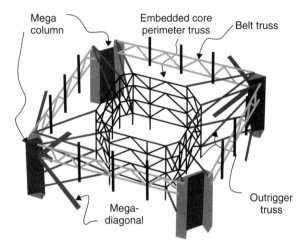

Figure 8.3 Important structural components (Courtesy of L.E. Robertson and S.T. See)

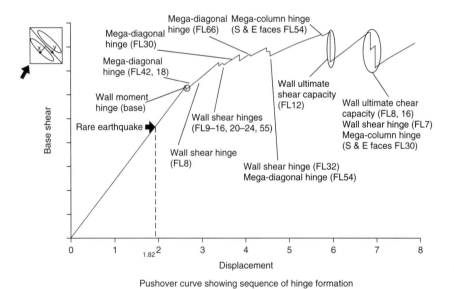

Pushover curve showing sequence of hinge formation

Figure 8.4 Static nonlinear analysis of the Shanghai World Financial Center (Courtesy of L.E. Robertson and S.T. See)

for a linear system. Using the step-by-step integration procedure, it is only necessary to reduce the effective stiffness matrix once and then back substitute the time dependent load vector for each time step. For the solution of a large system of simultaneous equations, the classic method of *Gauss Reduction* or one of its derivatives is a good approach to consider. It also has the advantage that it can be used for nonlinear response as well as linear response with only a slight modification. A brief summary of the basic method of Gauss Reduction follows.

8.3 GAUSS REDUCTION

One of the easier methods to obtain solutions for a system of n linear equations is by a process of elimination of unknowns. Of this family of possible solution methods, one of the more practical methods based on the idea of elimination is the Gauss Reduction method[2]. This method and its several variants form the basis for most techniques used in the solution of systems of linear equations. Consider the solution of a system of linear equations of the form:

$$a_{11}x_1 + a_{12}x_2 + \cdots\cdots + a_{1n}x_n = c_1$$
$$a_{21}x_1 + a_{22}x_2 + \cdots\cdots + a_{2n}x_n = c_2$$
$$\ldots\ldots\ldots\ldots\ldots\ldots\ldots\ldots\ldots\ldots\ldots\ldots\ldots \tag{8.7}$$
$$a_{n1}x_1 + a_{n2}x_2 + \cdots\cdots + a_{nn}x_n = c_n$$

We will seek a solution to a system of n linear equations of n unknowns x_1 to x_n. Dividing the first equation by a_{11} and solving for x_1 the result can be used to eliminate x_1 in the following equations. The resulting system of n-1 equations in $x_2, \ldots\ldots, x_n$ are obtained:

$$x_1 + a'_{12}x_2 + a'_{13}x_3 + \cdots\cdots + a'_{1n}x_n = c'_1$$
$$x_2 + a'_{23}x_3 + \cdots\cdots + a'_{2n}x_n = c'_2$$
$$\ldots\ldots\ldots\ldots\ldots\ldots\ldots\ldots\ldots \tag{8.8}$$
$$x_{n-1} + a'_{n-1,n}x_n = c'_{n-1}$$
$$x_n = c'_n$$

---

[2]Sokolnikoff, I. S. and Redheffer, R. M., *Mathematics of Physics and Engineering*, McGraw Hill Book Co. , 1958.

The substitution of $x_n = c'_n$ in the preceding equation in the set yields the value of x_{n-1} and by working backward in succession, the values of $x_{n-2}, x_{n-3}, \ldots, x_1$ are obtained.

By taking advantage of symmetry of the stiffness matrix, this can be accomplished computationally with a rather simple algorithm consisting of only 30 statements that include the following three major operations: (1) matrix reduction, (2) load vector reduction, and (3) back substitution. If there is no change in the stiffness for a time step, the matrix reduction is not required and only the load vector reduction and the back substitution are needed. When a new change in global stiffness occurs, all three steps are required, which results in increased computational time. MATLAB also has a variety of efficient built-in functions for solution of linear and nonlinear systems of equations.

In general, the stiffness matrices for very large structural systems contain relatively few nonzero terms. These types of matrices are called sparse or weakly populated. Specialized solution schemes have been developed for efficient solving of dynamic problems involving sparse matrices[3].

8.4 MATLAB APPLICATIONS

Filippou and Constantinides[4] have developed a MATLAB toolbox, called *FEDEASLab*, for linear and nonlinear, static and dynamic, structural analysis. This toolbox and its documentation, including several application examples, are available for download at no charge from http://fedeaslab.berkeley.edu.

The FEDEASLab toolbox consists of several functions that are grouped in categories and organized in separate directories. These functions operate on five basic data structures, which represent the model, the loading, the element properties, the state of the structural response, and the parameters of the solution strategy. To illustrate the utility of the FEDEASLab toolbox, consider the example application of the toolbox to static nonlinear analysis of the two-story frame shown in Figure 8.5. The frame is analyzed once ignoring second-order effects (i.e., P-Δ effects) and once including these effects. These two examples are part of the documentation of the toolbox.

---

[3]McGuire, W., Gallagher, R.H. and Ziemian, *R.D., Matrix Structural Analysis, 2nd Edition*, John Wiley & Sons, Inc. (2000).

[4]Filip C. Filippou and Margarita Constantinides, *FEDEASLab Getting Started Guide and Simulation Examples*, Technical Report NEESgrid-2004-22, University of California, Berkeley, 2004.

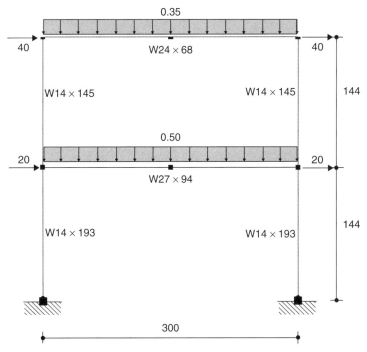

Figure 8.5 Geometry and loading of the two-story frame analyzed with the FEDEASLab MATLAB toolbox

The lateral loads shown in Figure 8.5 are the loads defining the load shape used for static nonlinear analysis (units shown are kips and inches).

The geometry of the frame is defined using the following MATLAB script:

```
%% ONE-BAY, TWO-STORY FRAME MODEL
%% Create Model
% all units in kips and inches
%% Node coordinates (in feet!)
XYZ(1,:) = [ 0     0]; % first node
XYZ(2,:) = [ 0    12]; % second node, etc
XYZ(3,:) = [ 0    24]; %
XYZ(4,:) = [25     0]; %
XYZ(5,:) = [25    12]; %
XYZ(6,:) = [25    24]; %
XYZ(7,:) = [12.5  12]; %
XYZ(8,:) = [12.5  24]; %
% convert coordinates to inches
XYZ = XYZ.*12;
```

```
%% Connectivity array
CON {1} = [  1    2];    % first story columns
CON {2} = [  4    5];
CON {3} = [  2    3];    % second story columns
CON {4} = [  5    6];
CON {5} = [  2    7];    % first floor girders
CON {6} = [  7    5];
CON {7} = [  3    8];    % second floor girders
CON {8} = [  8    6];
%% Boundary conditions
% (specify only restrained dof's)
BOUN(1,1:3) = [1 1 1];   % (1 = restrained,  0 = free)
BOUN(4,1:3) = [1 1 1];
%% Element type
% Note:  any 2 node 3dof/node element can be used at this point!
 [ElemName{1:8}] = deal('Lin2dFrm_NLG');
% 2d linear elastic frame element
%% Create model data structure
Model = Create_Model(XYZ,CON,BOUN,ElemName);
%% Display model and show node/element numbering
Create_Window (0.70,0.70);             % open figure window
Plot_Model  (Model);                   % plot model
Label_Model (Model);                   % label model
```

Execution of the preceding script on MATLAB creates the model and displays its geometry (Figure 8.6).

Nonlinear element properties are defined by the following script:

```
%% Define elements
% all units in kips and inches
%% Element name: 2d nonlinear frame element with concentrated
%% inelasticity
 [Model.ElemName{1:8}] = deal('OneCo2dFrm_NLG');
% One-component nonlinear 2d frame element
% [Model.ElemName{1:8}] = deal('EPPwHist2dFrm_NLG');
% Elastic-perfectly plastic nonlinear 2d frame element
% [Model.ElemName{1:8}] = deal('TwoCo2dFrm_NLG');
% Two-component nonlinear 2d frame element
%% Element properties
fy  = 50;          % yield strength
eta = 1.e-5;
% strain hardening modulus for multi-component models
%% Columns of first story W14x193
for i=1:2;
   ElemData{i}.E = 29000;
   ElemData{i}.A = 56.8;
   ElemData{i}.I = 2400;
   ElemData{i}.Mp  = 355*fy;
   ElemData{i}.eta = eta;
end
%% Columns of second story W14x145
for i=3:4;
   ElemData{i}.E = 29000;
```

```
    ElemData{i}.A = 42.7;
    ElemData{i}.I = 1710;
    ElemData{i}.Mp  = 260*fy;
    ElemData{i}.eta = eta;
end
%% Girders on first floor W27x94
for i=5:6;
    ElemData{i}.E = 29000;
    ElemData{i}.A = 27.7;
    ElemData{i}.I = 3270;
    ElemData{i}.Mp  = 278*fy;
    ElemData{i}.eta = eta;
end
%% Girders on second floor W24x68
for i=7:8;
    ElemData{i}.E = 29000;
    ElemData{i}.A = 20.1;
    ElemData{i}.I = 1830;
    ElemData{i}.Mp  = 177*fy;
    ElemData{i}.eta = eta;
end
%% Default values for missing element properties
ElemData = Structure ('chec',Model,ElemData);
```

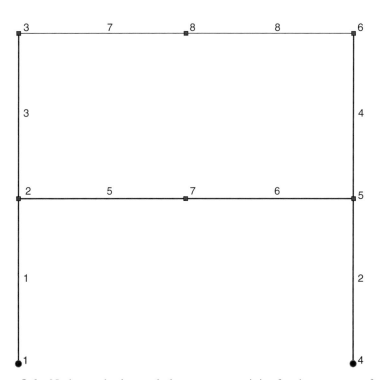

Figure 8.6   Node numbering and element connectivity for the two-story frame
analyzed with the FEDEASLab MATLAB toolbox

The script for nonlinear analysis with inclusion of geometric stiffness (P-Δ effects) is as follows:

```
%% TWO-STORY STEEL FRAME, PUSHOVER ANALYSIS WITH CONSTANT GRAVITY
%% LOADS AND LATERAL FORCES UNDER LOAD CONTROL for NONLINEAR
%% GEOMETRY
%====================================================================
%  FEDEASLab - Release 2.6, July 2004
%  Matlab Finite Elements for Design, Evaluation and Analysis
%  of Structures
%  Copyright(c) 1998-2004. The Regents of the University
%  of California. All Rights Reserved.
%  Created by Professor Filip C. Filippou
%  (filippou@ce.berkeley.edu)
%  Department of Civil and Environmental Engineering, UC Berkeley
%====================================================================
%% Initialization: clear memory and define global variables
% all units in kips and inches
CleanStart
%% Create output file
IOW = Create_File (mfilename);
%% Create Model
Model_TwoStoryFrm
% echo input data of structural model to output file (optional)
Print_Model (Model,'Pushover analysis of two story steel frame');
%% Element properties
SimpleNLElemData
% print element properties (optional)
Structure ('data',Model,ElemData);
%% 1. Loading (distributed loads and vertical forces on columns)
% define loading
for el=5:6 ElemData{el}.w = [0;-0.50]; end
for el=7:8 ElemData{el}.w = [0;-0.35]; end
Pe(2,2) =  -200;
Pe(3,2) =  -400;
Pe(5,2) =  -200;
Pe(6,2) =  -400;
GravLoading = Create_Loading (Model,Pe);
%% Nonlinear geometry option for columns
for el=1:4 ElemData{el}.Geom = 'PDelta'; end
%% Incremental analysis for distributed element loading
%% (single load step)
% initialize state
State = Initialize_State(Model,ElemData);
% initialize solution strategy parameters
SolStrat = Initialize_SolStrat;
% specify initial load increment (even though it is the same
% as the default value and could be omitted)
SolStrat.IncrStrat.Dlam0 = 1;
% initialize analysis sequence
 [State SolStrat] =
```

```
Initialize(Model,ElemData,GravLoading,State,SolStrat);
% apply load in one increment
 [State SolStrat] =
Increment(Model,ElemData,GravLoading,State,SolStrat);
% perform equilibrium iterations (we assume that convergence
% will occur!)
 [State SolStrat] =
Iterate   (Model,ElemData,GravLoading,State,SolStrat);
% update State
State = Update_State(Model,ElemData,State);
% determine resisting force vector
State   = Structure ('forc',Model,ElemData,State);
% set plot counter and store results for post-processing
k = 1;
Post(k) = Structure ('post',Model,ElemData,State);
%% 2. Loading in sequence: horizontal forces
% specify nodal forces
% !!!! IMPORTANT!!!! CLEAR PREVIOUS PE
clear Pe;
Pe(2,1) =   20;
Pe(3,1) =   40;
Pe(5,1) =   20;
Pe(6,1) =   40;

LatLoading = Create_Loading (Model,Pe);
%% Incremental analysis for horizontal force pattern (load control
%% is switched on)
%  (gravity forces are left on by not initializing State!)
%  specify initial load increment and turn load control on (default
%  value is ';no')
SolStrat.IncrStrat.Dlam0 = 0.40;
SolStrat.IncrStrat.LoadCtrl = ';yes';
SolStrat.IterStrat.LoadCtrl = ';yes';
% specify number of load steps
nostep = 20;
% initialize analysis sequence
[State SolStrat] =
Initialize(Model,ElemData,LatLoading,State,SolStrat);
tic;
% for specified number of steps, Increment, Iterate and
% Update_State (we assume again convergence!)
for j=1:nostep
   [State SolStrat] =
Increment(Model,ElemData,LatLoading,State,SolStrat);
   [State SolStrat] =
Iterate   (Model,ElemData,LatLoading,State,SolStrat);
   State = Update_State(Model,ElemData,State);
   k = k+1;
   Post(k) = Structure ('post',Model,ElemData,State);
end
toc;
%% Post-processing
```

```
% extract displacements from Post
np = length(Post);
x = zeros(np,1);
y = zeros(np,1);
pltDOF = Model.DOF(1,6);
supDOF = [Model.DOF(1,1) Model.DOF(1,4)];
for k=1:np
    x(k) = Post(k).U(pltDOF);
    y(k) = -sum(Post(k).Pr(supDOF))/120;
end
% plot force displacement relation in new window
fig = Create_Window(0.70,0.70);
ph1 = plot(x,y,'s-');
set (ph1,'MarkerSize',4,'MarkerFaceColor','b');
grid('on');
xlabel ('Horizontal roof displacement');
ylabel ('Load factor {\lambda}');
axis ([0 10 0 3]);
title ('Load factor-displacement');
% plot moment distribution at end of pushover
Create_Window(0.70,0.70);
Plot_Model(Model);
Plot_ForcDistr (Model,ElemData,Post(end),'Mz');
title('Moment distribution and plastic hinge locations near
        incipient collapse');
% show plastic hinge locations
Plot_PlasticHingeswPost(Model,Post(end));
% plot deformed shape of structure and plastic hinge locations
Create_Window(0.70,0.70);
Plot_Model(Model);
Structure('defo',Model,ElemData,State);

Plot_PlasticHingeswPost(Model,Post(end),State.U);
% close output file
fclose(IOW);
```

Upon execution of the script, the results of analysis are stored in a results file, and the moment diagram, the deformed shape at the end of time steps, and the force-displacement diagram are displayed (Figures 8.7 to 8.9).

Notice that if geometric stiffness were not considered, the force-displacement curve shown in Figure 8.10 would result, which would erroneously indicate that the capacity will never drop below that indicated by material strength.

FEDEASLab is also capable of performing nonlinear dynamic analysis. For example, Figure 8.11 shows the roof displacement of the two-story frame when the frame is subjected to the ground motion acceleration shown in Figure 8.12. Notice that the frame never comes back to its

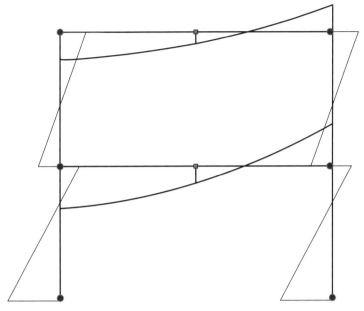

Figure 8.7 Moment distribution and plastic hinge locations at the end of pushover analysis

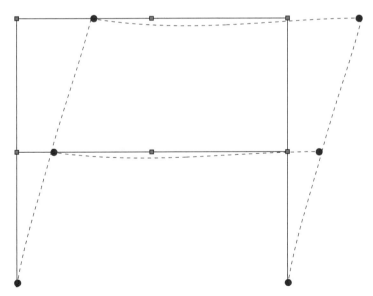

Figure 8.8 Displaced shape of the frame at the end of pushover analysis

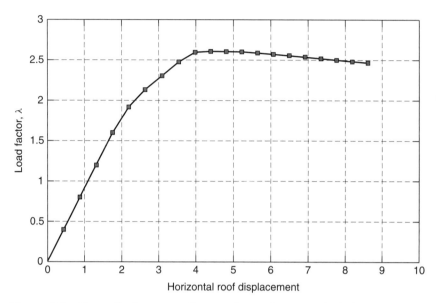

Figure 8.9   Force-displacement diagram for pushover analysis including geometric stiffness

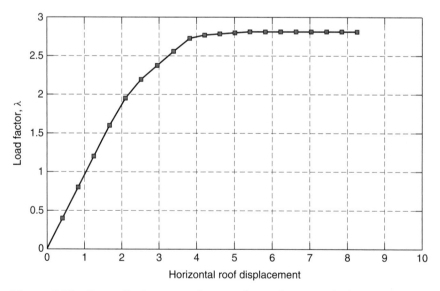

Figure 8.10   Force-displacement diagram for pushover analysis without geometric stiffness

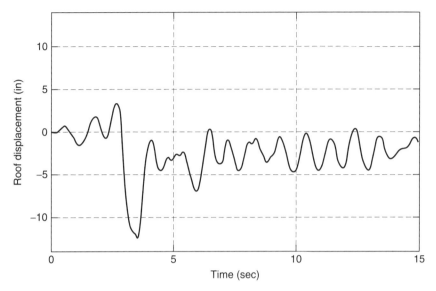

Figure 8.11   Dynamic roof displacement response of the two-story frame when subjected to the base acceleration history of Figure 8.12

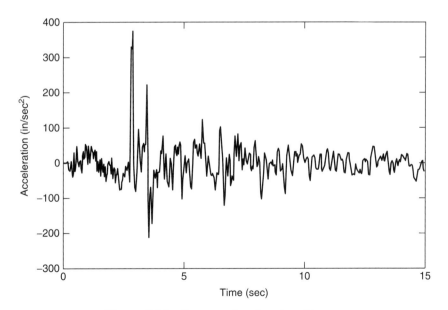

Figure 8.12   Base acceleration time history

initial position and exhibits significant permanent deformation, which is an indication of severe damage.

PROBLEMS

Problem 8.1(M)

Use the FEDEASLab MATLAB toolbox to model the six-story steel office building of Problem 3.1. Use this model to plot the mode shapes of the frame and calculate and plot the pushover response of the frame to a lateral load pattern consistent with the shape of its first mode. Perform the pushover analysis once ignoring the P-Δ effects and once including them. Are the pushover plots different and why?

CHAPTER 9

SEISMIC RESPONSE
OF STRUCTURES

9.1 INTRODUCTION

Structural dynamics is the study of structures in motion. Several natural and artificial events are capable of setting a structure in motion. Among these are wind, earthquake, air blast (explosion, pressure release), wave action, and vehicular generated motions. By far the most important of these to the structural engineer are the motions generated by the release of energy in an earthquake and their effects on the built environment. For this reason, in this chapter the field of earthquake engineering will be used to illustrate the application of the analysis techniques discussed in the previous chapters and to introduce the concepts of earthquake-resistant design.

   The increasing population and associated built environment in areas of high seismic activity has led to continued improvement in the development of analysis techniques and design procedures for earthquake resistance. The advent of high-speed digital computers and associated software has revolutionized the field of structural dynamics. During the 1970s, the development of the nuclear power–generating industry required that stringent seismic safety requirements be developed and used for the design of nuclear power stations built in the United States. This has led to a substantial increase in interest in and development of earthquake engineering, in particular, structural dynamics.

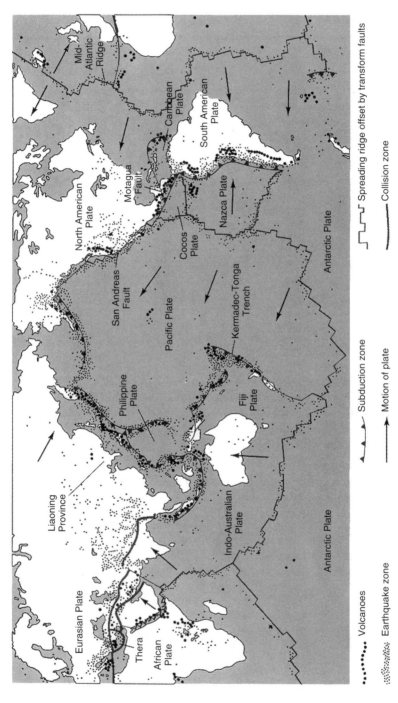

Figure 9.1 Plate tectonics and earthquakes

Volcanoes

Earthquake zone

Subduction zone

Motion of plate

Spreading ridge offset by transform faults

Collision zone

A general global explanation of the causes of earthquakes is closely related to the geological process called *plate tectonics*, which correlates well with the occurrence of earthquakes, as shown in Figure 9.1. The outermost layer of the earth, sometimes referred to as the *crust*, is broken into several large rock slabs called *plates*. Earthquakes tend to occur at the edges because of the movement of these plates relative to one another. The most significant zone of earthquake activity is known as the *circum-Pacific belt*, and, as the name implies, it surrounds the Pacific Ocean and includes four major plates. Earthquakes along the edges of this region affect countries located along its edges that include the following: the western edge of South America, Central America, California, and Alaska, then extending through the Aleutian Islands and the Kuril Islands to Japan and Taiwan, through the Philippines and Indonesia, and down to New Zealand.

9.2   LINEAR ELASTIC RESPONSE SPECTRA

As shown in Figure 9.2, there is considerable variation in the amplitude of the peak acceleration as well as differences in the frequency content and duration of earthquake ground motions. A means of evaluating the effect of these ground motion parameters on a particular structure is to pass the ground motion through a series of single-degree-of-freedom (SDOF) oscillators having different dynamic properties such as period and damping and calculating the resulting maximum response parameters. These parameters are determined as follows for an SDOF system.

Consider a damped oscillator subjected to a base acceleration. The equation of motion has the form

$$m\ddot{u} + c\dot{v} + kv = 0 \qquad (9.1)$$

where

$$\ddot{u} = \ddot{v}_g + \ddot{v}$$

With this substitution, the equation of motion has the form

$$m\ddot{v} + c\dot{v} + kv = -m\ddot{v}_g \qquad (9.2)$$

This equation can then be solved using either the Duhamel integral approach or step-by-step numerical integration. The Duhamel integral approach has the form

$$V(t) = \int_0^t \ddot{v}_g(\tau) \exp\left[-\xi\omega(t-\tau)\right] \sin\omega(t-\tau) d\tau \qquad (9.3)$$

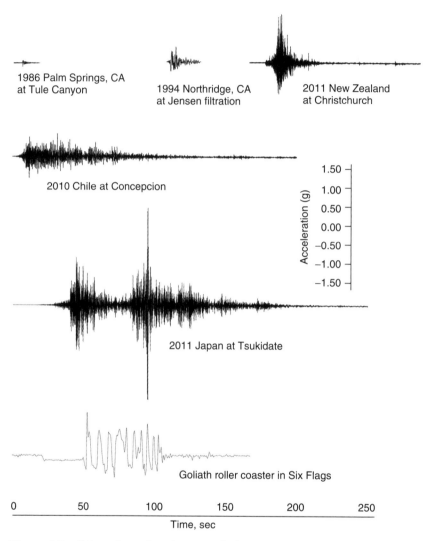

Figure 9.2 Selected acceleration records from a number of earthquakes and a roller coaster ride

where $V(t)$ is the earthquake response integral. Several step-by-step integration procedures have been discussed previously. To evaluate the earthquake response of a lumped MDOF system at any time requires evaluation of the earthquake response integral at that particular time for each significant response mode. This may require a substantial computational effort and makes the use of an approximate analysis based on the ground motion response spectra an attractive alternative.

The maximum response parameters can be determined as follows:

$$\ddot{u}_{max} = \left(\ddot{v} + \ddot{v}_g\right)_{max} = \text{maximum total acceleration}$$

$$\dot{v}_{max} = \text{maximum relative velocity} \tag{9.4}$$

$$v_{max} = \text{maximum relative displacement}$$

Define the following:

$$\text{SD} = S_d = |v_{max}| = \text{spectral displacement}$$

$$\text{SV} = S_v = |\dot{v}_{max}| = \text{spectral velocity} \tag{9.5}$$

$$\text{SA} = S_a = |\ddot{u}_{max}| = \text{spectral acceleration}$$

For the relatively small damping ratios found in most structures, the damping can be neglected in the following calculations. Consider the equation of motion for an oscillator without damping:

$$m\ddot{u} + kv = 0$$

$$\ddot{u} = -\frac{k}{m}v = -\omega^2 v \tag{9.6}$$

$$\ddot{u}_{max} = \omega^2 v_{max} = \omega^2 S_d$$

Consider the energy in an undamped oscillator. If there is no damping, the energy is conserved, and the energy balance equation can be expressed as

$$\text{total energy (TE)} = \text{kinetic energy (KE)} + \text{potential energy (PE)} \tag{9.7}$$

At the point of maximum displacement, the velocity is zero, and the total energy is equal to the potential energy. This can be expressed as

$$\text{TE} = \text{PE}_{max} = \frac{1}{2}kv_{max}^2 = \frac{1}{2}kS_d^2 \tag{9.8}$$

In a similar manner, at the point where the displacement crosses the horizontal axis, the displacement is zero, and the energy balance equation can be expressed as

$$\text{TE} = \text{KE}_{max} = \frac{1}{2}m\dot{v}_{max}^2 = \frac{1}{2}mS_v^2 \tag{9.9}$$

Because the damping is neglected, there is no energy dissipation, and the total energy is constant and

$$KE_{max} = PE_{max}$$

$$\frac{1}{2}mS_v^2 = \frac{1}{2}kS_d^2 \tag{9.10}$$

$$S_v = \omega S_d$$

Now, for a given ground acceleration record, calculate the time history displacement and determine

$$S_d = |v_{max}| = \text{spectral displacement} \tag{9.11}$$

Then determine

$$S_{pv} = \omega S_d = \text{spectral pseudovelocity}$$

$$S_{pa} = \omega^2 S_d = \text{spectral pseudoacceleration} \tag{9.12}$$

These two terms are given the prefix *pseudo* because they are not the true values of the maximum velocity and the maximum acceleration. However, for most strong-motion earthquakes, the spectral pseudoacceleration and true spectral acceleration are nearly identical. The use of the pseudovelocity and the pseudoacceleration permits the three response parameters to be plotted together on a sheet of three-way (tripartite) log paper. When combined with a plot of the maximum ground motion parameters, it is a good representation of the intensity of an earthquake ground motion. The only limitation occurs when the Duhamel integral is used to integrate the equation of motion because it is limited to elastic response. If a numerical integration procedure is used, it can readily be adapted to include nonlinear behavior.

The total energy of the structural system can be written as

$$E(t) = PE(t) + KE(t) = \frac{1}{2}k[v(t)]^2 + \frac{1}{2}m[\dot{v}(t)]^2 \tag{9.13}$$

Using Equation (9.12), we can construct the maximum energy response spectrum by evaluating the preceding equation for a range of frequencies and damping values. This energy is often expressed as

$$\sqrt{\frac{2E(t)}{m}} = \left\{[\omega v(t)]^2 + [\dot{v}(t)]^2\right\}^{1/2} \tag{9.14}$$

The tripartite response spectra for the ground motion recorded at El Centro, California, during the Imperial Valley earthquake of May 18, 1940, are shown in Figure 9.3 for the S00E component. This is one of the first strong-motion earthquake records and, as such, has been used for

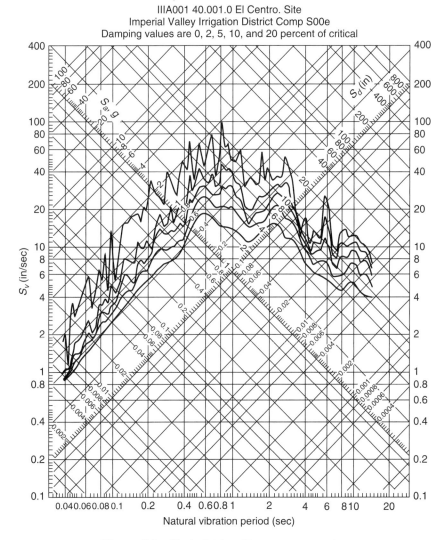

Figure 9.3 Typical tripartite response spectra

seismic analysis and design in many seismic regions of the world. Note that the time history record of this component is shown in Figure 9.2. In Figure 9.3, spectra are shown for damping values of 0, 2, 5, 10, and 20 percent of critical damping.

9.3 ELASTIC DESIGN RESPONSE SPECTRUM

The detailed characteristics of future earthquakes at a particular site are not known. Therefore, earthquake design spectra are obtained by averaging a set of response spectra from records of previous earthquakes having similar characteristics such as soil condition, epicentral distance,

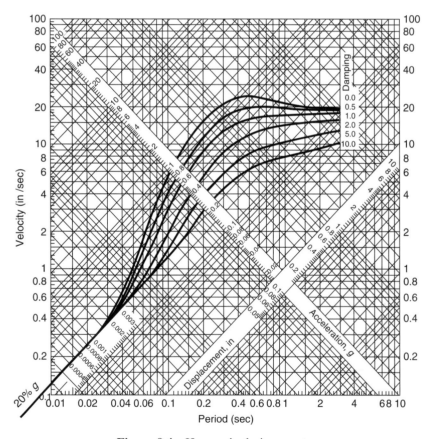

Figure 9.4 Housner's design spectra

magnitude, and source mechanism. Design spectra are presented as either smoothed curves or straight lines in order to eliminate peaks and valleys. Statistical analysis is generally used to create a smoothed spectrum at a suitable design level based on the seismic risk.

The first earthquake design spectrum was developed by Housner in 1959.[1] His smoothed spectra (Figure 9.4) were based on the characteristics of the two horizontal components of four earthquake ground motions recorded at El Centro, California (1934, 1940); Olympia, Washington (1949); and Taft, California (1970).

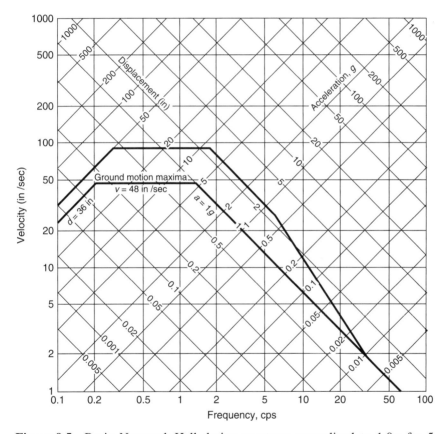

Figure 9.5   Basic Newmark-Hall design spectrum normalized to 1.0g for 5 percent damping

---

[1]G. W. Housner, "Behavior of Structures during Earthquakes," *Proc. ASCE*, Vol. 85 (EM4), October 1959.

Table 9.1 Relative Values of Spectrum Amplification Factors (after Newmark and Hall 1982)

Percentage of Critical Damping	Amplification Factor for		
	Displacement	Velocity	Acceleration
0	2.5	4.0	6.4
0.5	2.2	3.6	5.8
1	2.0	3.2	5.2
2	1.8	2.8	4.3
5	1.4	1.9	2.6
10	1.1	1.3	1.5
20	1.0	1.1	1.2

Newmark and Hall[2] recommended an earthquake design spectrum using straight lines to represent acceleration, velocity, and displacement, which tend to be constant in the high-, intermediate-, and low-frequency regions of the spectrum. The ground motion spectrum was normalized to a peak acceleration of $1.0g$, a peak velocity of 48 in/sec, a peak displacement of 36 in, and a damping of 5 percent. The spectrum for this normalized ground motion is shown in Figure 9.5. The amplified response to be used for a structure having a given period (frequency) is obtained by estimating the spectral amplification above the ground motion. Amplification factors based on the percentage of critical damping in the structure are given in Table 9.1. Control points are suggested at 33 Hz for the beginning point of the region of structural amplification, a transition region between 33 Hz and the maximum amplified acceleration region at 6.5 Hz. Two other control points occur: one at the intersection of the amplified velocity region and amplified acceleration region and the other at the intersection of the amplified displacement region and amplified velocity region. The two parameters needed to develop this design spectrum are the peak expected ground acceleration and the amount of structural damping. The ordinates of the design spectrum can then be scaled to the designed acceleration and damping.

Example 9.1 Construct a Newmark-Hall design spectrum for a peak acceleration of $0.3g$ and 5 percent critical damping (Table 9.1). Plot the design spectrum on a sheet of tripartite log paper, as shown in Figure 9.6.

---

[2]N. M. Newmark and W. J. Hall, "Procedures and Criteria for Earthquake Resistant Design," *Building Practices for Disaster Mitigation*, NBS, US Department of Commerce, Building Research Series, 46, 1973.

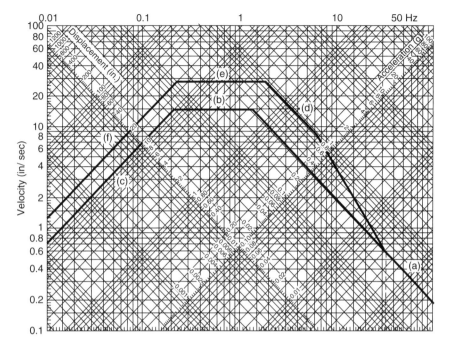

Figure 9.6

Ground Motion Parameters

$$\ddot{u} = 0.3 * 1.0 = 0.3g \qquad\qquad \text{(a)}$$

$$\dot{v} = 0.3 * 48 = 14.4 \text{ in/sec} \qquad \text{(b)}$$

$$v = 0.3 * 36 = 10.8 \text{ in} \qquad\qquad \text{(c)}$$

Amplified (Structure) Response Parameters

$$\ddot{u} = 2.6 * 0.3 = 0.78g \qquad\qquad \text{(d)}$$

$$\dot{v} = 1.9 * 14.4 = 27.4 \text{ in/sec} \qquad \text{(e)}$$

$$v = 1.4 * 10.8 = 19.4 \text{ in} \qquad\qquad \text{(f)}$$

Contemporary design spectra use probabilistic methods and rely on earthquake ground motion predictive relations also known as *attenuation relations*. A new generation of attenuation (NGA) relations forms the basis of design spectra adopted by the building codes in the United

States. An attenuation relation is a function relating various earthquake ground motion spectral parameters (spectral displacement, velocity, or acceleration) to a variety of factors such as earthquake magnitude, distance, site soil conditions, and faulting mechanisms. For example, the NGA relation developed by Campbell and Bozorgnia[3] has the following form:

$$\ln Y = f_{\text{mag}} + f_{\text{dis}} + f_{\text{flt}} + f_{\text{hng}} + f_{\text{site}} + f_{\text{sed}}$$

where Y is the median value of the spectral entity of interest and each function on the right side of the equation represents the effect on one or more earthquake characteristic. For example, f_{mag} is the magnitude term and has the following form:

$$f_{\text{mag}} = \begin{cases} c_0 + c_1 M & M \leq 5.5 \\ c_0 + c_1 M + c_2(M - 5.5) & 5.5 < M \leq 6.5 \\ c_0 + c_1 M + c_2(M - 5.5) + c_3(M - 6.5) & M > 6.5 \end{cases}$$

where M is the earthquake magnitude and c_0 to c_3 are constants determined based on regression analyses. Similarly, the distance term, f_{dis}, has the following form:

$$f_{\text{dis}} = (c_4 + c_5) \ln \left(\sqrt{R_{\text{rup}}^2 + c_6^2} \right)$$

where R_{rup} is the distance of the site from the earthquake rupture zone and c_4 to c_6 are constants determined based on regression analyses. Using attenuation relations such as the one introduced previously and probabilistic models of earthquake occurrence, we can develop site-specific design spectra corresponding to various return periods of earthquakes (Figure 9.7).

Building codes, however, select a few such relations, calculate spectral entities at a few distinct periods of vibration (usually 0.0, 0.20, and 1.0 sec), and construct a standard shape design spectrum that passes through these period points. Internet-based software for development of code design spectra at a site-based on-site location (in terms of either a zip code or latitude and longitude coordinates) is available for free and may be downloaded from the US Geological Survey (USGS) website.[4]

---

[3] K. W. Campbell and Y. Bozorgnia "NGA Ground Motion Model for the Geometric Mean Horizontal Component of PGA, PGV, PGD and 5% Damped Linear Elastic Response Spectra for Periods Ranging from 0.01 to 10 s," *Earthquake Spectra*, EERI, Vol. 24, No. 1, 2008.
[4] http://earthquake.usgs.gov/hazards/designmaps/javacalc.php.

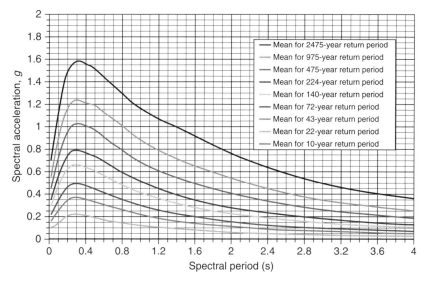

Figure 9.7   Uniform hazard spectra developed for a site in Los Angeles, California

Example 9.2   Use the USGS Hazard Calculator to develop a 2009 International Building Code (IBC) 5 percent damped design spectrum for a site with soil class D located at the zip code 90089.

Simply download and launch the USGS software. Select International Building Code as the analysis option and 2009 as the data edition and supply the zip code and site condition parameters. The design spectrum will be tabulated for you (see Figure 9.8). The same data are tabulated and also shown in graphical form in Figure 9.9.

Period (sec)	S_a (g)	S_d (in)
0.000	0.492	0.000
0.104	1.230	0.131
0.200	1.230	0.481
0.521	1.230	3.263
0.600	1.068	3.758
0.700	0.916	4.384
0.800	0.801	5.010
0.900	0.712	5.636

Period (sec)	S_a (g)	S_d (in)
1.000	0.641	6.263
1.100	0.583	6.889
1.200	0.534	7.515
1.300	0.493	8.142
1.400	0.458	8.768
1.500	0.427	9.394
1.600	0.401	10.020
1.700	0.377	10.647
1.800	0.356	11.273
1.900	0.337	11.899
2.000	0.321	12.525

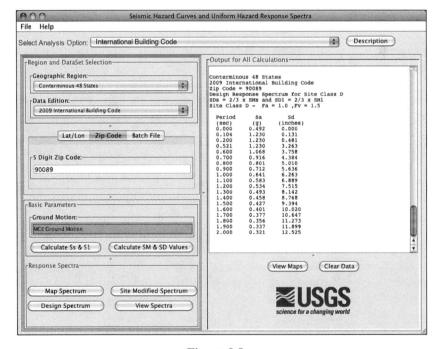

Figure 9.8

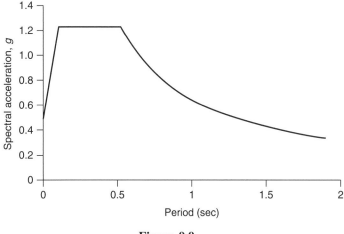

Figure 9.9

9.4 EARTHQUAKE RESPONSE OF SDOF SYSTEMS

When a single-story structure such as the one shown in Figure 2.2 is subjected to an earthquake ground motion, no external force is applied. Instead, the system experiences an acceleration at the base. The equation of motion for this structure becomes

$$m\ddot{u} + c\dot{v} + kv = 0 \qquad (9.15)$$

where $\ddot{u} = \ddot{v}_g + \ddot{v}$ is the total acceleration, which is the sum of the acceleration relative to the base plus the acceleration of the base. If the expression for the total acceleration is substituted into Equation (9.15), the equation of motion can be written in terms of the relative displacement:

$$m\ddot{v} + c\dot{v} + kv = -m\ddot{v}_g = p_{\text{eff}} \qquad (9.16)$$

For the purpose of this discussion, it is convenient to express the earthquake response in terms of the Duhamel integral although, in practice, it is more convenient to use one of the numerical integration procedures discussed earlier. Consider the Duhamel integral solution:

$$Y(t) = \frac{1}{m\omega_D} \int_0^t P(\tau) e^{-\xi\omega(t-\tau)} \sin \omega_D (t - \tau) d\tau \qquad (9.17)$$

The generalized force for a ground motion is

$$P_e^*(t) = -\ddot{v}_g \left[\int \mu(x)\phi(x)dx + \sum m_i \phi_i \right] = -\ddot{v}_g \zeta \qquad (9.18)$$

where ζ is the earthquake excitation factor representing the extent to which the earthquake motion tends to excite response in the assumed deflected shape. Substituting the generalized equivalent force into the integral, we obtain

$$Y(t) = \int_0^t \frac{\zeta \ddot{v}_g(\tau)}{m^* \omega_D} e^{-\xi \omega(t-\tau)} \sin \omega_D (t - \tau) d\tau \qquad (9.19)$$

Now define

$$V(t) = \int_0^t \ddot{v}_g(\tau) e^{-\xi \omega(t-\tau)} \sin \omega_D (t - \tau) d\tau \qquad (9.20)$$

Then

$$Y(t) = \frac{\zeta}{m^* \omega_D} V(t) \qquad (9.21)$$

where $[V(t)]_{max} = S_v =$ the spectral velocity

The maximum relative displacement becomes

$$v(x)_{max} = \phi(x) \frac{\zeta}{m^* \omega_D} S_v = \phi(x) \frac{\zeta}{m^*} S_d \qquad (9.22)$$

The acceleration can be written in terms of the generalized coordinate amplitude as

$$\ddot{Y}(t) = \omega_D^2 Y(t) \qquad (9.23)$$

For a system with distributed mass, the inertia force is

$$q(x,t) = \mu(x)\ddot{v}(x,t) = \mu(x)\phi(x)\ddot{Y}(t) \qquad (9.24)$$

And the maximum inertia force can be expressed as

$$[q(x)]_{\max} = \mu(x)\phi(x)\omega_D^2 [Y(t)]_{\max}$$

$$[q(x)]_{\max} = \mu(x)\phi(x)\omega_D^2 \frac{\zeta}{m^*\omega_D} S_v \tag{9.25}$$

$$[q(x)]_{\max} = \mu(x)\phi(x)\frac{\zeta}{m^*} S_a$$

An important parameter for the seismic design of buildings is the shear force at the base, which can be expressed by summing the inertia forces:

$$Q(t) = \text{base shear} = \int_0^h q(x,t)dx$$

$$Q_{\max} = \int_0^h q(x)_{\max}dx = \frac{\zeta}{m^*}S_a \int_0^h \mu(x)\phi(x)dx = \frac{\zeta^2}{m^*}S_a \tag{9.26}$$

where h = the height of the building.

For the purposes of design, the spectral acceleration is usually specified, and Equation (9.26) can be used to determine the design base shear. The base shear can then be distributed over the height of the building in the following manner:

$$q(x) = \mu(x)\phi(x)\frac{\zeta}{m^*}\left(\frac{Q_{\max}m^*}{\zeta^2}\right) = \mu(x)\phi(x)\frac{Q_{\max}}{\zeta} \tag{9.27}$$

This is the process generally used in the building codes to define the design criteria for buildings. For a lumped-mass system, Equation (9.26) can be written as

$$q_i = \frac{m_i\phi_i}{\zeta}Q_{\max} = m_i\phi_i\frac{\zeta}{m^*}S_a \tag{9.28}$$

$$Q_{\max} = \frac{\zeta^2}{m^*}S_a = \frac{[\sum w_i\phi_i]^2 g}{g^2 \sum w_i\phi_i^2}S_a = W^*\frac{S_a}{g} = CW^* \tag{9.29}$$

where C = a code-specified seismic coefficient based on the percentage of gravity, g

W^* = the effective weight

Most codes, however, take the total dead weight as the effective weight.

The overturning moment at the base can also be calculated as

$$M_o = \sum h_i q_i$$

where $i = 1, \ldots, n =$ the number of stories

Equation (9.26) can also be expressed in terms of the spectral acceleration as

$$S_a = \frac{Q_{\max} m^*}{\zeta^2} \tag{9.30}$$

This form of Equation (9.26) can be applied to evaluate the strength of older existing buildings, where the shear strength is known based on the lateral force criteria used for the design and the spectral acceleration capacity can be estimated.

Example 9.3   In Example 7.2, based on the static deformed shape, the first mode of vibration for the four-story steel building was calculated with a corresponding period of 0.702 sec. Assume 5 percent damping and calculate the base shear and overturning moment for the four-story building for the following parameters of the 1940 El Centro earthquake.

Level	W_i	ϕ	$W_i \phi_i$	$W_i \phi_i^2$
4	294	1.00	294.0	294.0
3	368	0.902	331.9	299.4
2	368	0.706	259.8	183.4
1	370	0.470	173.9	81.7
Total	1400		1059.6	858.6

$$W^* = \frac{\left(\sum w_i \phi_i \right)^2}{\sum w_i \phi_i^2} = \frac{(1059.6)^2}{858.6} \doteq 1307.7$$

For the 1940 Imperial Valley (El Centro) earthquake, N–S:

$$T = 0.70 \text{ sec} \quad S_a = 0.60g \quad S_v = \frac{S_a}{\omega} = \frac{0.6(386.4)}{8.95} = 25.9$$

$$\zeta = \sum m_i \phi_i$$

$$= \frac{1}{386.4}[294.0(1) + 368(0.902) + 368(0.706 + 370(0.470)] = 2.742$$

$$Q_{max} = \frac{\zeta^2}{m^*}S_a = \frac{(2.742)^2}{2.22}(0.6)(386.4) = 785.2 \text{ kips}$$

$$q_{max} = \frac{m_i \phi_i}{\zeta}Q_{max}$$

Level	$\dfrac{m_i \phi_i}{\zeta}$	q	h_i
4	0.278	218.0	43.5
3	0.313	245.8	33.0
2	0.245	192.4	22.5
1	0.164	129.0	12.0

$$M_o = \sum q_i h_i = 218(43.5) + 245.8(33) + 192.4(22.5) + 129(12)$$

$$M_o = 23,445 \text{ ft-kips} \ (281,340 \text{ in-kips})$$

$$Y_{max} = \frac{\zeta}{m^*\omega} \qquad S_v = \frac{2.742}{2.22(8.95)}(25.9) = 3.57 \text{ in}$$

$$v(x)_{max} = \phi(x)Y_{max}$$

$$v(x)_{max} = \begin{Bmatrix} 3.57 \\ 3.22 \\ 2.52 \\ 1.68 \end{Bmatrix} \text{ in}$$

9.5   EARTHQUAKE RESPONSE ANALYSIS OF MDOF SYSTEMS

For analyzing the earthquake response of linear systems, it is generally convenient to transform to a system of normal (modal) coordinates because the structural response to the support motions is primarily due to the lower modes. Therefore, a good approximation for the response of a system containing many degrees of freedom can be obtained by performing an analysis for only a few normal coordinates. This transformation to normal coordinates results in an equation of motion for the nth mode of the form

$$M_n \ddot{Y}_n + C_n \dot{Y}_n + K_n Y_n = P_n(t) \tag{9.31}$$

where M_n, C_n, and $K_n =$ the generalized properties of the nth mode
$Y =$ the amplitude of the nth mode response

By analogy to the generalized SDOF systems, the response of the nth mode can be determined as

$$Y_n(t) = \frac{\zeta_n}{M_n^* \omega_n} V_n(t) \tag{9.32}$$

or the maximum response can be expressed in terms of the spectral displacement as

$$(Y_n)_{\max} = \frac{\zeta_n}{M_n^* \omega_n} S_{vn} = \frac{\zeta_n}{M_n^*} S_{dn} \tag{9.33}$$

The relative displacement produced by the nth mode can be written as

$$\{v_n(t)\} = \{\phi_n\} Y_n(t) = \{\phi_n\} \frac{\zeta_n}{M_n^* \omega_n} V_n(t) \tag{9.34}$$

The maximum relative displacement is given in terms of the spectral displacement as

$$(v_n)_{\max} = \{\phi_n\} \frac{\zeta_n}{M_n^*} S_{dn} = \{\phi_n\} \alpha_n S_{dn} \quad \alpha_n = \frac{\zeta_n}{M_n^*} \tag{9.35}$$

The acceleration produced by the nth mode becomes

$$\ddot{Y}_n(t) = \omega_n^2 Y_n(t) = \frac{\zeta_n}{M_n^*} \omega_n V_n(t) \tag{9.36}$$

and the maximum acceleration in this mode is given in terms of the spectral acceleration as

$$(\ddot{Y}_n)_{\max} = \frac{\zeta_n}{M_n^*} \omega_n S_{vn} = \frac{\zeta_n}{M_n^*} S_{an} = \alpha_n S_{an} \tag{9.37}$$

In geometric coordinates, the maximum acceleration has the form

$$(\ddot{v}_n)_{\max} = \{\phi_n\} (\ddot{Y}_n)_{\max} = \{\phi_n\} \alpha_n S_{an} \tag{9.38}$$

The maximum lateral force produced by the nth mode can be expressed in terms of the mass and spectral acceleration as

$$\{q_n\}_{\max} = [M]\{\ddot{v}_n\}_{\max} = [M]\{\phi_n\}\alpha_n S_{an} \tag{9.39}$$

Summing these forces over the structure results in the shear force at the base for the nth mode:

$$Q_n = \sum_{i=1}^{N} q_{in} = \{I\}^T[M]\{\phi_n\}\alpha_n S_{an} = \frac{\zeta_n^2}{M_n^*} S_{an} \tag{9.40}$$

$$Q_n = W_n \frac{S_{an}}{g}$$

where $W_n = \dfrac{\zeta_n^2}{M_n^*} g = $ the effective weight of the nth mode

Multiplying the forces for the nth mode by the height of each force above the base produces the moment at the base:

$$M_{on} = \{h\}^T (q_n)_{\max} = \{h\}^T[M]\{\phi_n\}\alpha_n S_{an} \tag{9.41}$$

Evaluation of the earthquake response of an MDOF system at any point in time requires evaluation of the modal response for each time increment during the earthquake time history.

9.5.1 Time History Modal Analysis

The normal equation for each mode can also be written in terms of the generalized properties as

$$M_n^* \ddot{Y}_n + C_n^* \dot{Y}_n + K_n^* Y = P_n^*(t) \tag{9.42}$$

where

$$P_n^*(t) = -\{\phi\}_n^T[M]\{1\}\ddot{v}_g(t) \tag{9.43}$$

It will be necessary to integrate each modal equation as for an SDOF system and combine directly the response for each mode at each time step (instant of time). Neglecting damping, we can write Equation (9.43) as

$$\ddot{Y}_n + \omega_n^2 Y_n = -\frac{\{\phi_n\}^T[M]\{1\}}{M_n^*}\ddot{v}_g = -\Gamma_n \ddot{v}_g(t) \tag{9.44}$$

where the modal participation factor is defined as

$$\Gamma_n = \frac{\{\phi_n\}^T [M]\{1\}}{\{\phi_n\}^T [M]\{\phi_n\}} \tag{9.45}$$

9.5.2 Modal Combinations for Spectral Analyses

The modal response quantities obtained from a spectral analysis are modal maxima that do not necessarily occur at the same instant of time. Therefore, to get an estimate of the actual maximum at a specific point in time, some method of combination must be used. Several methods have been proposed, including the following:

1. Simple sum (SS): $r \simeq \sum\limits_{n=1}^{N} r_n$

2. Sum of absolute values (SAV): $r \leq \sum\limits_{n=1}^{N} |r_n|$

3. Square root of sum of squares (SRSS): $r \simeq \sqrt{\sum\limits_{n=1}^{N} r^2}$

4. Complete quadratic combination (CQC): $r \simeq \sqrt{\sum\limits_{i=1}^{N} \sum\limits_{j=1}^{N} r_i \rho_{ij} r_j}$

where

$$\rho_{ij} = \frac{8\xi^2(1+\lambda)\lambda^{3/2}}{(1-\lambda^2)^2 + 4\lambda\xi^2(1+\lambda)^2} \qquad \xi = \frac{c}{c_{cr}} = \text{constant} \qquad \lambda = \frac{\omega_j}{\omega_i}$$

Example 9.4   Repeat Example 9.3 for the four-story steel frame but consider all four modes of vibration.

Structural Properties

$$[K] = 500 \begin{bmatrix} 1.175 & -1.175 & 0 & 0 \\ -1.175 & 2.351 & -1.175 & 0 \\ 0 & -1.175 & 2.643 & -1.467 \\ 0 & 0 & -1.467 & 2.451 \end{bmatrix}$$

$$[M] = \begin{bmatrix} 0.761 & & & \\ & 0.952 & & \\ & & 0.952 & \\ & & & 0.958 \end{bmatrix}$$

Vibration Properties

$$\Phi = \begin{bmatrix} 1.000 & 1.000 & 1.000 & 1.000 \\ 0.897 & 0.188 & -1.048 & -2.251 \\ 0.678 & -0.815 & -0.389 & 3.143 \\ 0.432 & -0.956 & 0.944 & -2.204 \end{bmatrix}$$

$$\{\omega\} = \begin{Bmatrix} 8.95 \\ 25.03 \\ 39.93 \\ 48.68 \end{Bmatrix} \qquad \{T\} = \begin{Bmatrix} 0.702 \\ 0.251 \\ 0.157 \\ 0.129 \end{Bmatrix}$$

Spectral Values

See Figure 9.10.

$$\{S_{vn}\} = \begin{Bmatrix} 26.0 \\ 17.0 \\ 7.5 \\ 6.0 \end{Bmatrix} \frac{\text{in}}{\text{sec}} \qquad \{S_{an}\} = \begin{Bmatrix} 0.60 \\ 0.83 \\ 0.53 \\ 0.69 \end{Bmatrix} g \qquad \{S_{dn}\} = \begin{Bmatrix} 2.80 \\ 0.55 \\ 0.20 \\ 0.13 \end{Bmatrix} \text{in}$$

Generalized Properties

$$\zeta_n = \{1\}^T [M]\{\phi_n\} = \begin{Bmatrix} 2.67 \\ -0.75 \\ 0.30 \\ -0.50 \end{Bmatrix} \qquad \alpha_n = \frac{\zeta_n}{M_n^*} = \begin{Bmatrix} 1.25 \\ -0.33 \\ 0.11 \\ -0.03 \end{Bmatrix}$$

$$[M^*] = [\phi]^T [M][\phi] = \begin{bmatrix} 2.143 & 0 & 0 & 0 \\ 0 & 2.303 & 0 & 0 \\ 0 & 0 & 2.804 & 0 \\ 0 & 0 & 0 & 19.64 \end{bmatrix}$$

Overturning Moment

$$\begin{Bmatrix} M_{o1} \\ M_{o2} \\ M_{o3} \\ M_{o4} \end{Bmatrix} = \begin{bmatrix} x_1 & 0 & 0 & 0 \\ x_1 + x_2 & x_2 & 0 & 0 \\ x_1 + x_2 + x_3 & x_2 + x_3 & x_3 & 0 \\ x_1 + x_2 + x_3 + x_4 & x_2 + x_3 + x_4 & x_3 + x_4 & x_4 \end{bmatrix} \begin{Bmatrix} q_1 \\ q_2 \\ q_3 \\ q_4 \end{Bmatrix}$$

where x measured from the top of the structure.

Response Quantities

See Tables 9.2 through 9.6.

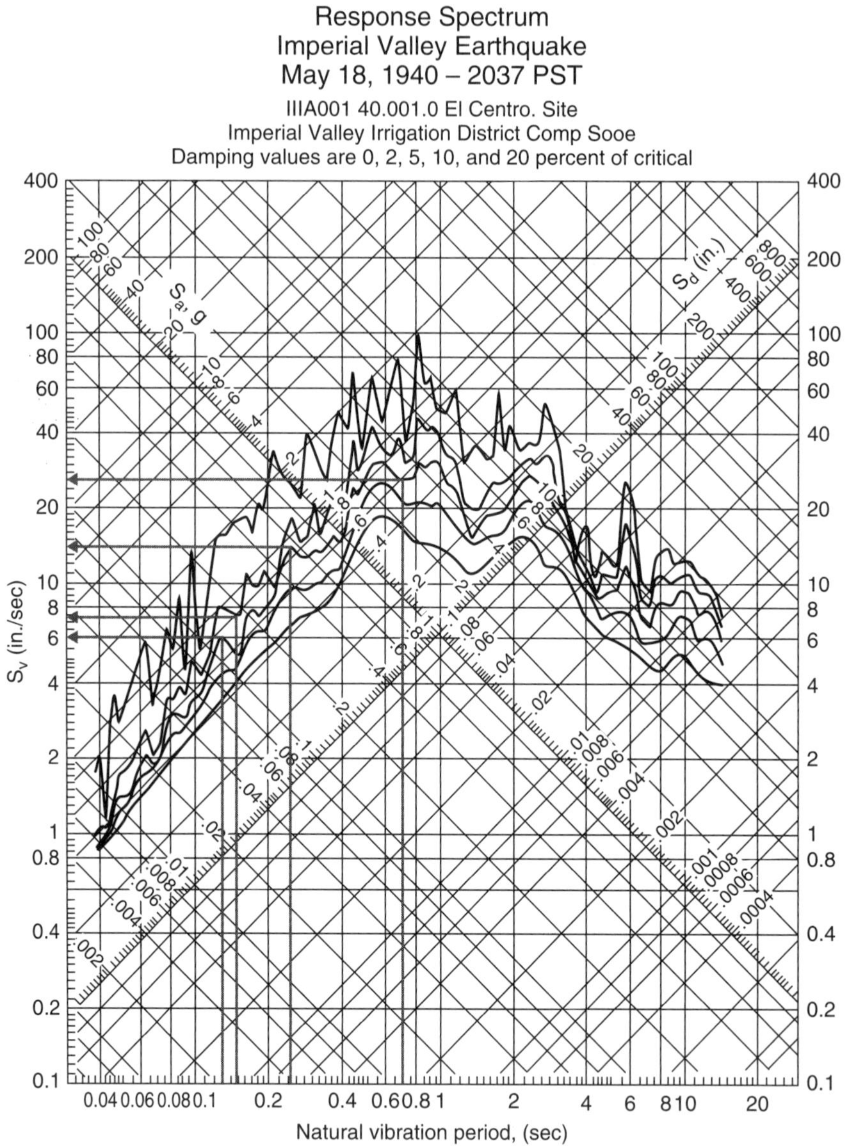
Response Spectrum
Imperial Valley Earthquake
May 18, 1940 – 2037 PST
IIIA001 40.001.0 El Centro. Site
Imperial Valley Irrigation District Comp Sooe
Damping values are 0, 2, 5, 10, and 20 percent of critical
S_v (in./sec)
S_a, g
S_d (in.)
Natural vibration period, (sec)

Figure 9.10

Table 9.2   Displacement: $\{v_n\}_{\max} = \{\phi_n\}Y_{n\,\max} = \{\phi_n\}\alpha_n S_{dn}$

	Modal Response				Combined Response	
	1	2	3	4	SAV	SRSS
$N = 4$	3.50	−0.18	0.02	0.004	3.704	3.505
$N = 3$	3.15	−0.03	−0.02	−0.009	3.209	3.150
$N = 2$	2.38	0.15	−0.01	0.012	2.417	2.385
$N = 1$	1.51	0.17	0.02	−0.009	1.709	1.520

Table 9.3   Acceleration: $\{\ddot{v}_n\} = \omega_n^2\{v_n\}$

	Modal Response				Combined Response	
	1	2	3	4	SAV	SRSS
$N = 4$	280.3	−112.7	31.89	9.5	434.4	303.9
$N = 3$	252.3	−18.8	−31.89	−21.3	324.3	255.9
$N = 2$	190.6	94.0	15.94	28.4	328.9	215.0
$N = 1$	121.0	106.5	31.89	−21.3	280.6	165.7

Table 9.4   Inertia Force: $\{q_n\} = [M]\{\ddot{v}_n\}$

	Modal Response				Combined Response	
	1	2	3	4	SAV	SRSS
$N = 4$	213.1	−85.8	24.3	−7.2	549.1	231.1
$N = 3$	240.2	−17.9	−30.4	20.3	308.8	243.6
$N = 2$	181.4	60.9	−15.2	−27.0	284.5	193.8
$N = 1$	115.9	102.0	88.0	20.4	326.3	178.9

Table 9.5   Shear: $Q_n = \sum\limits_{i=1}^{N} q_{in}$

	Modal Response				Combined Response	
	1	2	3	4	SAV	SRSS
$N = 4$	213.1	−85.8	24.3	−7.2	330.4	231.1
$N = 3$	453.3	−103.7	−6.1	13.1	576.2	456.2
$N = 2$	634.7	−42.8	−21.3	−13.9	712.7	636.6
$N = 1$	750.6	59.2	66.7	6.5	883.0	755.9

Table 9.6 Overturning Moment: $M_{on} = \sum\limits_{i=1}^{N} h_i q_{in}$

	Modal Response				Combined Response	
	1	**2**	**3**	**4**	**SAV**	**SRSS**
$N = 4$	2,238	−901	255	76	3,470	2,427
$N = 3$	6,378	−1,990	446	62	8,876	6,696
$N = 2$	13,662	−2,439	414	−98	16,613	13,884
$N = 1$	22,669	−1,729	1,177	−6.4	25,581	22,765

Example 9.5(M)   Repeat Example 9.4 using MATLAB to calculate the following:

a. Modal displacements, story forces, and base shears
b. SRSS and CQC values of the displacements, story forces, and base shears

We first define the spectral values and store them in Sa and Sd arrays:

```
%
% Define spectral values
%
Sa=[0.60, 0.83, 0.53, 0.69];
Sd=[2.80, 0.55, 0.20, 0.13];
```

In order to have a general script that we can use for a shear building of any number of stories and the desired number of modes with little modification, we define the number of stories, NS, and the number of modes, NM, as variables:

```
NM=4;
NS=4;
```

Now we are ready to calculate the mode shapes and frequencies using MATLAB's eig function. We also define the unity vector, Kapa, which we will need later in the process:

```
%
% Set up mass and stiffness matrices and call eig
%
g=386.4;
zeta = 0.05;
```

```
%
K=500.*[1.175 -1.175 0 0; -1.175 2.351 -1.175 0; 0 -1.175 2.643
        -1.467; 0 0 -1.467 2.451];
%
M=[0.761 0 0 0; 0 0.952 0 0; 0 0 0.952 0; 0 0 0 0.958] ;
%
% Unity Vector
%
Kapa = ones(NS,1);
%
[ModeShapes, Eigenvalues] = eig(K,M)
```

MATLAB displays the computed eigenvalues and eigenvectors:

```
ModeShapes =

    -0.6832    -0.6592     0.5930    -0.2473
    -0.6125    -0.1241    -0.6292     0.5139
    -0.4630     0.5369    -0.2299    -0.7035
    -0.2956     0.6300     0.5644     0.4909


Eigenvalues =

    1.0e+03 *

    0.0799         0          0          0
         0    0.6267          0          0
         0         0     1.5910          0
         0         0          0     2.3765
```

Next we calculate the natural frequencies and periods and normalize the mode shapes:

```
%
% natural frquencies and periods
%clc
for i = 1:NM
    Omega (i,i) = sqrt(Eigenvalues(i,i));
    T(i) = 2*pi()/Omega(i,i);
end
%
% Normalize modes
%
for i=1:NM
    ModeShapes(:,i)=ModeShapes(:,i)./ModeShapes(1,i) ;
end
%
% Display periods and normalized mode shapes
%
T, ModeShapes
```

MATLAB will display the results as

```
T =

    0.7029     0.2510      0.1575      0.1289

ModeShapes =

    1.0000      1.0000      1.0000      1.0000
    0.8965      0.1882     -1.0609     -2.0784
    0.6777     -0.8145     -0.3876      2.8453
    0.4327     -0.9557      0.9518     -1.9853
```

a. Modal displacements, story forces, and base shears:

```
%
% Generalized mass
%
Mstar=ModeShapes'*M*ModeShapes
%
% Modal Participation
%
MP= ModeShapes'*M*Kapa
%
% Calculate modal values
%
for i=1:NM
      alpha(i) = MP(i)/Mstar(i,i);
    % Modal displacements
      v(:,i) = ModeShapes(:,i).*alpha(i)*Sd(i);
    % Modal accelerations
      a(:,i) = ModeShapes(:,i).*alpha(i)*Sa(i);
    % Modal story forces
      f(:,i) = M.*g*a(:,i);
    % Modal base shears
      Q(:,i) = f(:,i)'*Kapa;
end
%
% Display modal values
%
alpha, v, a, f, Q
```

MATLAB will display the following results:

```
v =
      3.4944     -0.1794      0.0207     -0.0033
      3.1328     -0.0338     -0.0219      0.0068
      2.3682      0.1462     -0.0080     -0.0093
      1.5119      0.1715      0.0197      0.0065
```

```
a =
    0.7488    -0.2708     0.0548    -0.0173
    0.6713    -0.0510    -0.0581     0.0360
    0.5075     0.2206    -0.0212    -0.0493
    0.3240     0.2588     0.0521     0.0344


f =
  220.1865   -79.6240    16.1076    -5.0965
  246.9436   -18.7497   -21.3771    13.2509
  186.6765    81.1337    -7.8099   -18.1404
  119.9261    95.7988    19.2995    12.7375


Q =
  773.7326    78.5588     6.2201     2.7515
```

b. SRSS and CQC values of the displacements, story forces, and base shears:

```
%
% SRSS Values
%
for n=1:NS
    vSRSS(n,:) = sqrt( v(n,:)* v(n,:)');
    fSRSS(n,:) = sqrt( f(n,:)* f(n,:)');
end
    QSRSS = sqrt( Q*Q');
%
% Display SRSS values
%
vSRSS, fSRSS, QSRSS
%
% CQC Coefficients
%
for i=1:NM
    for j=1:NM
        Lambda = Omega(j,j)/Omega(i,i);
        Rho(i,j)= 8*(zeta^2)*(1+Lambda)*Lambda^1.5;
        Rho(i,j)= Rho(i,j)/((1-Lambda^2)^2
                 + (4*Lambda*zeta^2)*(1+Lambda)^2);
    end
end
%
% CQC values (easy way)
%
for n=1:NS
    vCQC(n,:) = sqrt( v(n,:)*Rho* v(n,:)');
    fCQC(n,:) = sqrt( f(n,:)*Rho* f(n,:)');
end
    QCQC = sqrt( Q*Rho*Q');
%
```

```
% Display CQC values
%
vCQC,  fCQC,QCQC
```

MATLAB will display the results:

```
vSRSS =
     3.4991
     3.1331
     2.3728
     1.5217

fSRSS =
   234.7499
   248.9282
   204.5015
   155.2237

QSRSS =
   777.7403

vCQC =
     3.4977
     3.1327
     2.3738
     1.5232

fCQC =
   233.9546
   248.5760
   204.8678
   156.8089

QCQC =
   778.3890
```

To plot the results, we use the following simple commands:

```
%
% Plot the results
%
Yscale =NM:-1:0;
Mplot = [ModeShapes;zeros(1,NM)]
Vplot = [v;zeros(1,NM)];
fplot = [f;zeros(1,NM)];
vplotSRSS = [vSRSS;0];
fplotSRSS = [fSRSS;0];
```

```
vplotCQC = [vCQC;0];
fplotCQC = [fCQC;0];
%
figure (1);
subplot(1,3,1);
plot(Mplot,Yscale);
subplot(1,3,2);
plot(Vplot,Yscale);
subplot(1,3,3);
plot(fplot,Yscale);
%
figure (2);
subplot(2,2,1);
plot(vplotSRSS,Yscale);
subplot(2,2,2);
plot(fplotSRSS,Yscale);
subplot(2,2,3);
plot(vplotCQC,Yscale);
subplot(2,2,4);
plot(fplotCQC,Yscale);
```

MATLAB will display the results in graphic form, as shown in Figures 9.11 and 9.12.

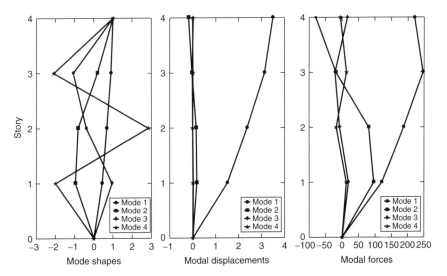

Figure 9.11

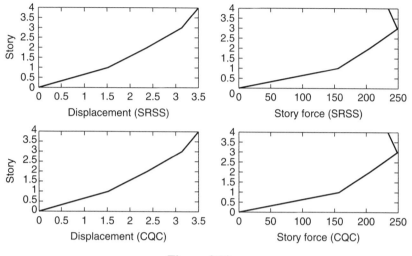

Figure 9.12

9.6   STRUCTURAL DYNAMICS IN THE BUILDING CODE

The predominant national model building code in the United States is the International Building Code, generally referred to as the 2009 IBC (International Code Council 2009). Regional and local codes usually adopt the national code with some amendments. For example, the California Building Code (CBC) is an amended version of the IBC, whereas the Los Angeles Building Code is an amended version of the CBC. The 2009 IBC with some minor exceptions adopts the provisions of the ASCE 7 standard for earthquake effects by reference. As stated in the 2009 IBC, "every structure ... shall be designed and constructed to resist the effects of earthquake motions in accordance with ASCE 7 ...." The edition of ASCE 7 referenced in the 2009 IBC is ASCE 7-05.[5] However, the 2010 edition of ASCE 7 has already been published (ASCE 2010), which will be referenced in future editions of the IBC.

ASCE 7 defines spectral accelerations based on the USGS mapped acceleration parameters similar to what was discussed in Example 9.2. The design spectral accelerations at two periods of vibration, 0.2 sec (S_{DS}) and 1.0 sec (S_{D1}), are used to anchor the shape of the design spectrum.

[5]American Society of Civil Engineers, *Minimum Design Loads for Buildings and Other Structures*, ASCE/SEI Standard 7-05, 2005.

The site soil conditions are factored into the establishment of the S_{DS} and S_{D1} values. The design spectrum shape is defined as

$$0 \leq T \leq T_0 = \frac{0.2 S_{D1}}{S_{DS}} \rightarrow S_a = S_{DS}\left(0.4 + 0.6\frac{T}{T_0}\right)$$

$$T_0 < T \leq T_S = \frac{S_{D1}}{S_{DS}} \rightarrow S_a = S_{DS}$$

$$T_S < T \leq T_L \rightarrow S_a = \frac{S_{D1}}{T} \qquad (9.46)$$

$$T > T_L \rightarrow S_a = \frac{S_{D1} T_L}{T^2}$$

where T_L is the long-period transition period shown on the relevant maps. T_L is equal to or larger than 6.0 sec in most areas of the United States and therefore rarely controls design of building structures.

The resulting design spectrum shape is shown in Figure 9.13.

Based on the function of the building and the number of occupants, the building is assigned an *occupancy category* and a corresponding *importance factor*, I, which varies from 1.0 for ordinary buildings to 1.5 for essential facilities. Depending on the occupancy category and values of S_{DS} and S_{D1} at the site, each building is assigned a *seismic design category*, which varies from A to E. An additional category F is

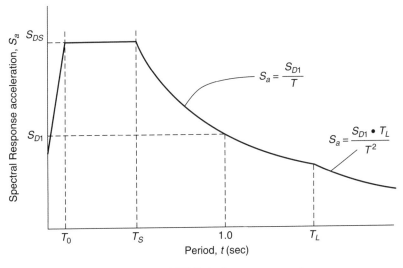

Figure 9.13 ASCE 7 design spectrum shape

assigned to buildings constructed on very poor site conditions such as liquefiable soils. As one moves from A to F, the seismic analysis and design provisions become more complex and elaborate. For example, all California buildings are in category D or higher.

Based on past performance and experimental results, a *response modification factor*, R, is assigned to each structural system. Larger values of R correspond to structural systems believed to be more ductile. For example, the value of R for a special moment-resisting frame (steel or reinforced concrete) is 8.0, whereas the value of R for an ordinary moment frame varies from 3.0 to 3.5. Similarly, each structural system is assigned a *system overstrength factor*, Ω_0, and a *deflection amplification factor*, C_d. Elastic forces are divided by R to reflect reduced design forces for the structural system, which is expected to be ductile and behave in a nonlinear fashion and therefore will experience smaller forces compared to an equivalent elastic system. Such reduced forces, however, underestimate the actual lateral displacements of the system because the nonlinear system will deform as much if not more than the equivalent elastic system. The deflection amplification factor is intended to correct this discrepancy.

Values of Ω_0 vary from 1.25 to 3.0. Values of C_d vary from 1.25 to 6.0. For example, for both steel and reinforced concrete special moment frames, $\Omega_0 = 3.0$ and $C_d = 5.5$. In addition, certain height limits are assigned to a number of structural systems in certain seismic design categories. For example, special reinforced concrete shear wall systems cannot be taller than 160 ft in seismic design categories D and E and no taller than 100 ft in seismic design category F.

Depending on the structure's seismic design category, structural system, dynamic properties, and regularity, or lack thereof, either the *equivalent lateral force procedure* or the *modal response spectrum procedure* is most often used for establishing the response parameters of a structure. The use of nonlinear dynamic analysis, however, is on the rise particularly in performance-based assessment of existing structures or design of new tall or special structures.

9.6.1 Equivalent Lateral Force Procedure

For this procedure, the seismic base shear is calculated from

$$V = C_S W \tag{9.47}$$

where C_S = the seismic response coefficient
W = the effective seismic weight of the structure

C_S is calculated as follows:

$$C_S = \frac{S_{DS}}{\left(\dfrac{R}{I}\right)} \tag{9.48}$$

C_S calculated per Equation (9.48) need not exceed the following values:

$$C_S = \frac{S_{D1}}{T\left(\dfrac{R}{I}\right)} \quad \text{for } T \le T_L \tag{9.49}$$

$$C_S = \frac{S_{D1}T_L}{T^2\left(\dfrac{R}{I}\right)} \quad \text{or } T > T_L \tag{9.50}$$

However, C_S should not be less than

$$C_S = 0.044 S_{DS} I \ge 0.01 \tag{9.51}$$

In addition, for structures located where S_1 (a spectral parameter obtained from the hazard map or the USGS Hazard Calculator) is equal to or greater than $0.6g$, C_S should not be less than

$$C_S = \frac{0.5 S_1}{\left(\dfrac{R}{I}\right)} \tag{9.52}$$

Equations (9.48) to (9.50) simply define C_S as spectral acceleration, as defined in Equation (9.46) and shown in Figure 9.13 with two adjustments: (1) the rising branch of spectral acceleration for $0 \le T \le T_0$ is conservatively replaced with a constant spectral acceleration that covers the entire range of $0 \le T \le T_S$, and (2) the values of elastic spectral acceleration are divided by (R/I) to represent reduced forces because of the nonlinear response and presumed ductility of the system. Therefore, Equation (9.47) represents the familiar concept of equating the base shear of an SDOF system to its effective weight times the spectral acceleration at its fundamental period of vibration.

The fundamental period of the structure may be calculated either from an approximate formula (Method A) or from rational analysis, interpreted as the use of the Rayleigh method or from a more complex eigenvalue or computer analysis (Method B).

The Method A estimate of the fundamental period, T_a, is obtained from

$$T_a = C_t h_n^x \tag{9.53}$$

where h_n = the height, in feet, above the base to the highest level of the structure

C_t, x = the tabulated structural system-dependent parameters

For example, for steel moment-resisting frames, $C_t = 0.028$ and $x = 0.8$. For concrete moment-resisting frames, $C_t = 0.016$ and $x = 0.9$. ASCE 7 also provides a couple of alternative formulas for calculating T_a.

An upper-bound limit is placed on the calculated fundamental period (Method B):

$$T \leq C_u T_a \tag{9.54}$$

where C_u = a coefficient that depends on the S_{D1} value for the site and varies from 1.4 for $S_{D1} \geq 0.4$ to 1.7 for $S_{D1} \leq 0.1$

The story forces are calculated as follows:

$$F_x = C_{vx} V \tag{9.55}$$

$$C_{vx} = \frac{w_x h_x^k}{\sum_{i=1}^{n} w_i h_i^k} \tag{9.56}$$

where w_i, w_x = the portion of the total effective seismic weight of the structure, W, located or assigned to level i or x

h_i, h_x = the height from the base to level i or x

$$k = \begin{cases} 1.0 & T \leq 0.5 \text{ sec} \\ 2.0 & T \geq 2.0 \text{ sec} \\ \text{Interpolate} & 0.5 < T < 2.0 \text{ sec} \end{cases}$$

Once the story forces are calculated, the story shears can be obtained by summing the story forces:

$$V_x = \sum_{i=x}^{n} F_i \tag{9.57}$$

where F_i = the portion of the seismic base shear, V, induced at level i

The design deflections at any level are calculated by amplifying the displacements obtained from elastic analysis for reduced forces, δ_{xe}:

$$\delta_x = \frac{C_d \delta_{xe}}{I} \qquad (9.58)$$

It should be clear from the preceding discussion that the equivalent lateral force procedure is nothing but a simple application of generalized coordinates to reduce an MDOF system to an equivalent SDOF system. Equation (9.56) represents an implicitly assumed deformed shape for the equivalent system. Incorporation of k makes this shape adjustable from linear ($k = 1$) to parabolic ($k = 2$).

9.6.2 Modal Response Spectrum Procedure

This procedure requires an analysis to determine the natural modes of vibration for the structure similar to what we have already covered in this chapter for MDOF systems. A sufficient number of modes should be considered to obtain a minimum of 90 percent modal mass participation.

The value of forces for each mode of response obtained using the design spectrum is multiplied by I and divided by R to obtain the design modal forces. The values for displacement and drift quantities are multiplied by C_d and divided by I to obtain modal design displacements.

Modal values are combined using either the SRSS or the CQC method. The CQC method should be used for closely spaced modes that have significant cross-correlation of translational and torsional response.

The combined base shear, V_t, is then compared to the base shear calculated from application of the equivalent lateral force procedure, V. If $V_t < 0.85V$, then the forces, but not the displacements, are scaled upward by multiplying them by

$$0.85 \frac{V}{V_t}.$$

Example 9.6(M) Repeat Example 9.5(M) using the code design spectrum developed in Example 9.2. Compare your results with those obtained in Example 9.5(M).

In engineering practice, it is rare to know the exact value of the natural periods of vibration of a structure prior to performing some analyses. Therefore, it is nice to have a function in which the design spectrum is defined in its entirety. Furthermore, as analysis progresses and the natural periods are calculated, a simple interpolation function calculates the spectral values at the desired periods of vibration. In order to do this, we modify the script used in Example 9.5(M) as follows.

First, replace the following commands:

```
%
% Define spectral values
%
Sa=[0.60, 0.83, 0.53, 0.69];
Sd=[2.80, 0.55, 0.20, 0.13];
```

with

```
Periods= [0.000, 0.104, 0.200, 0.521, 0.600, 0.700, 0.800, 0.900,
          1.000, 1.100, 1.200, 1.300, 1.400, 1.500, 1.600, 1.700,
          1.800, 1.900, 2.000];
%
SAvals = [0.492, 1.230, 1.230, 1.230, 1.068, 0.916, 0.801, 0.712,
          0.641, 0.583, 0.534, 0.493, 0.458, 0.427, 0.401, 0.377,
          0.356, 0.337, 0.321];
%
SDvals =[0.000, 0.131, 0.481, 3.263, 3.758, 4.384, 5.010, 5.636,
          6.263, 6.889, 7.515, 8.142, 8.768, 9.394, 10.020, 10.647,
          11.273, 11.899, 12.525];
```

Now we can use MATLAB's interpolation function, interp1, to obtain spectral ordinates at any given period. For example, to get spectral acceleration at $T = 1.05$ sec, we simply use a statement such as

```
SaTest = interp1(Periods, SAvals,1.05)
```

Next, insert the following interpolation commands immediately before calculating the modal values:

```
%
% Interpolate to obtain spectral acceleration and displacement
% for each mode
%
for i=1:NM
    Sa(i) = interp1(Periods, SAvals,T(i));
    Sd(i) = interp1(Periods, SDvals,T(i));
end
%
% Calculate modal values
```

We are now ready to execute the script, which will display the following results [the periods and mode shapes will be the same as those in Example 9.5(M) and therefore are not repeated]:

```
MP =
    2.6741
   -0.7508
    0.2939
   -0.4108

alpha =
    1.2480   -0.3262    0.1034   -0.0251

v =
    5.4942   -0.3011    0.0337   -0.0056
    4.9256   -0.0567   -0.0358    0.0116
    3.7235    0.2452   -0.0131   -0.0158
    2.3771    0.2878    0.0321    0.0111

a =
    1.1390   -0.4013    0.1271   -0.0309
    1.0211   -0.0755   -0.1349    0.0642
    0.7719    0.3269   -0.0493   -0.0879
    0.4928    0.3835    0.1210    0.0613

f =
  334.9113 -117.9971   37.3818   -9.0850
  375.6097  -27.7857  -49.6109   23.6212
  283.9414  120.2342  -18.1249  -32.3372
  182.4117  141.9669   44.7893   22.7059

Q =
    1.0e+03 *
    1.1769    0.1164    0.0144    0.0049

vSRSS =
    5.5026
    4.9261
    3.7316
    2.3947

fSRSS =
  357.1678
  380.6231
  310.5691
  236.5381

QSRSS =
    1.1827e+03

vCQC =
    5.5003
    4.9256
    3.7334
    2.3972
```

```
fCQC =
  355.7663
  379.8313
  311.1166
  239.7321

QCQC =
  1.1837e+03
```

As expected, the SRSS and CQC results are virtually identical. The CQC combined forces and displacements obtained in the two examples are compared in Figures 9.14 and 9.15, which demonstrate that the design spectrum defined in Example 9.2 imposes substantially more demand on the structure than the 1940 El Centro response spectrum.

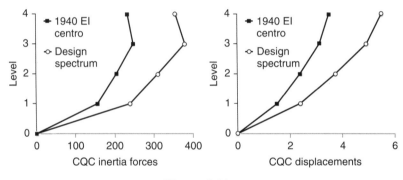

Figure 9.14

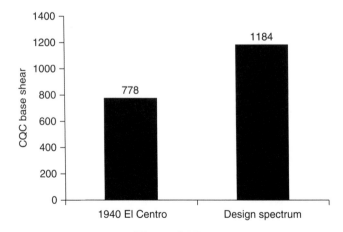

Figure 9.15

Example 9.7 For the four-story steel frame of Example 9.6(M), given a site with soil class D, $R = 8$, $I = 1$, $C_d = 5.5$, $S_S = 1.85$, $S_1 = 0.64$, $T_L = 6.0$ sec, and the design spectrum of the same example, determine:

a. ASCE 7 design story forces and base shears using the equivalent lateral force procedure
b. ASCE 7 design story forces, design base shears, and design displacements using the modal response spectrum procedure considering four modes of vibration
c. The number of modes necessary to satisfy the 90 percent mass participation requirement

a. Equivalent Lateral Force Procedure

From previous examples, $T = 0.70$ sec based on rational analysis:

$$T_a = C_t h_n^x = (0.028)(43.5)^{0.8} = 0.57 \text{ sec}$$

$$1.4T_a = 1.4(0.57) = 0.80 > T = 0.70$$

Therefore, we can use T.

$$C_S = \frac{S_{DS}}{\left(\dfrac{R}{I}\right)} = \frac{1.23}{\left(\dfrac{8}{1}\right)} = 0.154$$

Check C_S against the upper- and lower-bound limits:

$$T = 0.70 < T_L = 6.0$$

$$C_S \leq \frac{S_{D1}}{T\left(\dfrac{R}{I}\right)} = \frac{0.641}{0.70\left(\dfrac{8}{1}\right)} = 0.114 \text{ (controls)}$$

$$0.044 S_{DS} I = 0.044(1.23)(1.0) = 0.054 \geq 0.01 \text{ (does not control)}$$

$$S_1 = 0.64 > 0.60$$

$$S_1 = 0.64 > 0.6 \rightarrow C_S \geq \frac{0.5 S_1}{\left(\dfrac{R}{I}\right)}$$

$$= \frac{0.5(0.64)}{\left(\dfrac{8}{1}\right)} = 0.040 \text{ (does not control)}$$

$$V = C_S W = 0.114(1400) = 159.60 \text{ kips}$$

Interpolate for k:

$$k = \frac{2(T+1)}{3} = \frac{2(0.70+1)}{3} = 1.133$$

Now calculate the story forces, F_x, and story shears, V_x, using Equations (9.55) to (9.57):

Level	W_x	h_x	h_x^k	$W_x h_x^k$	C_{vx}	F_x	V_x
4	294	43.50	71.85	21,123.00	0.36	56.98	56.98
3	368	33.00	52.54	19,334.11	0.33	52.16	109.14
2	368	22.50	34.04	12,527.68	0.21	33.79	142.93
1	370	12.00	16.70	61,78.94	0.10	16.67	159.60
Total				59,163.73		159.60	

b. Modal Response Spectrum Procedure

We can start with either the SRSS or the CQC results from Example 9.6(M). We use the CQC values in this example.

The CQC base shear from Example 9.6(M) = 1184 kips.

$$V_t = 1184 \left(\frac{1.0}{8.0} \right) = 148 \text{ kips} > 0.85(159.60) = 135.66$$

Therefore, there is no need to adjust V_t upward.

To obtain the design forces, we multiply the forces obtained in Example 9.6(M) by $I = 1$ and divide them by $R = 8$ (see Table 9.7).

To obtain the design displacements, we divide the displacements obtained in Example 9.6(M) by $R = 8$ and multiply them by $C_d = 5.5$ (the I values cancel each other out; see Table 9.8).

Table 9.7   Design Story Forces (kips)

Level	CQC Force from Example 9.6(M)	ASCE 7 Design Force
4	355.7663	44.47
3	379.8313	47.48
2	311.1166	38.89
1	239.7321	29.97

Table 9.8   Design Displacements (in)

Level	CQC Displacements from Example 9.6(M)	ASCE 7 Design Displacements
4	3.4977	2.40
3	3.1327	2.15
2	2.3738	1.63
1	1.5232	1.05

It can be seen that the design forces obtained by this procedure are smaller than those obtained using the equivalent lateral force procedure. This is because the building codes want to provide an incentive for engineers to use the more complex dynamic procedures instead of the simplified static procedures.

c. Mass Participation Factors

The modal participation factors were defined by Equation (9.45) as

$$\Gamma_n = \frac{\{\phi_n\}^T [M]\{1\}}{\{\phi_n\}^T [M]\{\phi_n\}}$$

The mass participation factors are obtained by dividing Γ_n by the total mass of the structure.

We can calculate the mass participation factors by simply inserting the following commands to the end of our MATLAB script for Example 9.6(M):

```
%
%   Mass Participation Factors
%
Ln = (MP.*MP)./(diag(Mstar).*diag(Mstar));
TotalMass = diag(M)'*ones(NS,1);
MPFACTORS = Ln./ TotalMass
```

Executing the script results in the display of the mass participation factors:

```
MPFACTORS =

    0.9212
    0.0676
    0.0084
    0.0028
```

This shows that 92 percent of the mass participation for this frame comes from mode 1 followed by only 6 percent from mode 2, 0.08 percent from mode 3, and 0.028 percent from mode 4. Therefore, considering only one mode satisfies the 90 percent participation requirement of ASCE 7. Note, however, that for many tall or unusual three-dimensional structures, dozens of modes must be considered to reach this level of mass participation.

PROBLEMS

Problem 9.1

Consider the six-story steel office building of Problem 3.1 and model it as a system having six degrees of freedom represented by the horizontal displacement at each floor level. The calculated mode shapes of the first two modes of vibration are given below along with the corresponding frequencies:

$$\omega = \begin{Bmatrix} 5.2 \\ 14.6 \end{Bmatrix} \text{rad/sec} \qquad [\phi] = \begin{bmatrix} 1.00 & 1.00 \\ 0.92 & 0.56 \\ 0.76 & -0.17 \\ 0.63 & -0.64 \\ 0.46 & -0.87 \\ 0.33 & -0.80 \end{bmatrix}$$

Using the elastic response spectra for the ground motion shown in Figure 9.16, determine the dynamic response considering the first two modes. Assume the damping is 5 percent of critical for both modes. Calculate story shears and displacement envelopes for both modes and combine the modal results using SAV and SRSS.

Problem 9.2(M)

Use MATLAB to solve Problem 9.1. Ignore the natural frequencies and mode shapes given in Problem 9.1 and use the values you obtain from MATLAB. In addition to the SAV and SRSS values, calculate the CQC values of response and compare your results to those obtained in Problem 9.1.

Problem 9.3(M)

Repeat Problem 9.2(M), considering six modes of vibration instead of only two. Compare your results to those obtained in Problem 9.2(M).

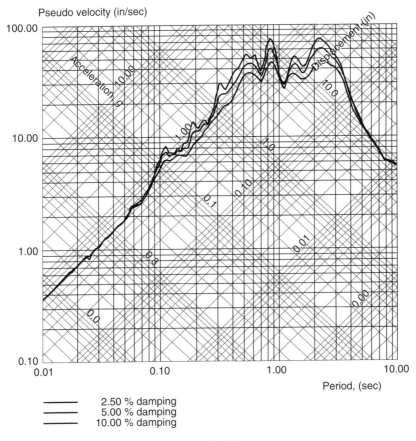

2.50 % damping
5.00 % damping
10.00 % damping

Figure 9.16

Problem 9.4(M)

Repeat Problem 9.3(M) but this time use (a) 2.5 percent damping and (b) 10 percent damping. Compare your results to those obtained in Problem 9.3(M). How significant or insignificant is the value of damping assumed?

Problem 9.5

Consider the six-story steel office building of Problem 3.1 and assume $\xi = 5\%$. Develop an SDOF dynamic model considering the normalized deflected shape (mode shape) and the lumped masses shown below. The linear elastic responsespectrum for the ground motion recorded at the Nordhoff Fire Station during the 1994 Northridge earthquake has the following characteristics at $T = 1.2$ sec: $S_a = 0.4g, S_v = 30$ in/sec, and $S_d = 5.8$ in.

a. Develop and plot a Newmark-Hall design spectrum for the same damping and same base acceleration as the recorded data.

b. Determine the maximum base shear for the earthquake and compare this value to that obtained from the design spectrum.

c. Plot the distribution of inertia forces and story shears over the height of the building for both spectra.

$$\langle\phi\rangle = \langle 1.00 \quad 0.915 \quad 0.762 \quad 0.625 \quad 0.460 \quad 0.331\rangle$$
$$\langle m\rangle = \langle 1.56 \quad 1.66 \quad 1.66 \quad 1.66 \quad 1.66 \quad 1.90\rangle$$
$$\langle k\rangle = \langle 583 \quad 583 \quad 874.5 \quad 874.5 \quad 1248.5 \quad 511.8\rangle$$
$$\omega = \frac{2\pi}{T} \qquad T = 1.2 \text{ sec}$$

Problem 9.6(M)

Use MATLAB to solve Problem 9.5.

Problem 9.7

A Los Angeles parking structure is instrumented by the California Strong Motion Instrumentation Program (CSMIP).[6] Sensors (accelerometers) are installed throughout this building to record the response of the building during earthquakes. The layout of the instrumentation of the building is shown in Figure 9.17. (The sensors are numbered and their directions are indicated.)

Assume the following mass matrix and dynamic properties for the first three modes of vibration of the building in the east–west direction, 5 percent critical damping, and the response spectrum of the ground motion shown in Figure 9.3 to determine the following using the SRSS method to combine modal responses, if required:

a. The displacement envelope over the height of the building considering the first mode of vibration.

b. The distribution of inertia forces over the height of the building considering only the first mode.

[6]The data and recorded motions for this building are available at http://strongmotioncenter.org.

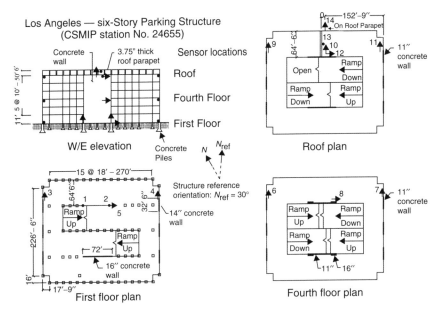

Figure 9.17

c. The base shear considering three modes of vibration.

d. The effective weight of the structure considering the first mode.

$$T = \begin{Bmatrix} 0.366 \\ 0.091 \\ 0.045 \end{Bmatrix} \sec$$

$$[\phi] = \begin{bmatrix} 0.143 & 0.109 & 0.096 \\ 0.114 & 0.244 & -0.036 \\ 0.084 & -0.057 & -0.105 \\ 0.056 & -0.106 & -0.033 \\ 0.031 & -0.107 & 0.088 \\ 0.011 & -0.622 & 0.105 \end{bmatrix}$$

$$[M] = \begin{bmatrix} 16.9 & & & & & \\ & 22.0 & & & & \\ & & 22.0 & & & \\ & & & 22.0 & & \\ & & & & 23.4 & \\ & & & & & 23.4 \end{bmatrix}$$

Problem 9.8(M)

Use MATLAB to solve Problem 9.7.

Problem 9.9(M)

During the 1994 Northridge earthquake, sensor 5, located at the ground level of the building of Problem 9.7, recorded a motion with the following 5 percent damped spectral characteristics. Repeat Problem 9.8(M) for this response spectrum.

```
Periods (T):
 .040    .042    .044    .046    .048    .050    .055    .060
 .065    .070    .075    .080    .085    .090    .095    .100
 .110    .120    .130    .140    .150    .160    .170    .180
 .190    .200    .220    .240    .260    .280    .300    .320
 .340    .360    .380    .400    .420    .440    .460    .480
 .500    .550    .600    .650    .700    .750    .800    .850
 .900    .950   1.000   1.100   1.200   1.300   1.400   1.500
1.600   1.700   1.800   1.900   2.000   2.200   2.400   2.600
2.800   3.000   3.200   3.400   3.600   3.800   4.000   4.200
4.400   4.600   4.800   5.000   5.500   6.000   6.500   7.000
```

```
Spectral Accelerations (g):
0.164   0.163   0.167   0.170   0.167   0.159   0.160   0.162
0.157   0.171   0.178   0.198   0.233   0.249   0.241   0.210
0.251   0.252   0.264   0.321   0.381   0.416   0.492   0.554
0.613   0.606   0.392   0.325   0.361   0.372   0.356   0.296
0.282   0.277   0.266   0.277   0.305   0.300   0.293   0.262
0.242   0.253   0.251   0.329   0.362   0.309   0.309   0.309
0.260   0.218   0.216   0.192   0.152   0.130   0.127   0.138
0.138   0.136   0.121   0.096   0.077   0.061   0.056   0.051
0.035   0.028   0.023   0.021   0.021   0.027   0.028   0.024
0.017   0.013   0.012   0.011   0.009   0.007   0.006   0.005
```

Problem 9.10(M)

Repeat Problem 9.9(M), using the code design spectrum given in Example 9.2 instead of the spectrum recorded by sensor 5 during the 1994 Northridge earthquake. Elaborate on the observed differences.

APPENDIX — HISTORICAL DEVELOPMENT OF BUILDING CODE SEISMIC PROVISIONS

Buildings that experience earthquake ground motion at the base will move with the ground and vibrate. To be successful, the design criteria and the resulting design must be able to produce a structure that will adequately resist any earthquakes that will affect the building during its intended useful life, including, in some cases, multiple earthquakes. The building code specifies the minimum amount of strength that will be required to satisfy specific risk criteria. Traditionally, most building codes have required that structures be designed to resist static lateral forces, which are a percentage of the vertical self-weight of the building. Vertical seismic forces are generally not included because it is generally thought that static gravity loads plus normal factors of safety should be adequate to resist the vertical effects of the earthquake motion. However, details must be carefully designed to provide for stress reversals and inelastic deformations (ductility). In addition, connections must be able to provide the necessary overstrength.

If the previous conditions are satisfied, the structure should be able to withstand minor earthquakes with limited nonstructural damage and without structural damage, moderate earthquakes without extensive structural damage, and major earthquakes without collapse. A well-designed structure should also minimize economic loss in the event of a moderate or strong earthquake. This requires careful attention to such things as material properties, connection details, and workmanship, which cannot be explicitly required in the code. The following brief history of building codes provides background for the development of these codes as well as the building code requirements for older buildings that may still be in use and in need of strengthening.

A.1   HISTORICAL OVERVIEW

Among the early regulations were those developed in Italy in 1909. These were based on the performance of wood-framed buildings that experienced the earthquake of 1908. A commission ruled that buildings be designed to resist a lateral force of 1/12 (0.083) of their self-weight. In 1912, the commission ruled that the first story should be designed for 1/12 of the weight above and that the second and third stories should be designed for 1/8 of the weight above. At that time, three-story buildings were the tallest permitted.

Following the San Francisco earthquake of 1906, which, combined with the resulting fires, destroyed most of the city, the city was rebuilt under code provisions that required a wind load of 30 lb/sq ft. The lateral force resulting from this loading was intended to provide resistance to seismic forces as well as wind forces, and no consideration was given to earthquake forces. In addition, the code did not require consideration of the mass of the structure.

Following the 1923 Kanto earthquake in Japan, which destroyed large parts of Tokyo and Yokohama, the seismic design law was adopted officially in 1924. It required that a design seismic coefficient of $0.1g$ or more should be used for all important new structures.

In 1927, the Uniform Building Code (UBC) was first enacted by the International Conference of Building Officials (ICBO). The seismic provisions were "suggested for inclusion in the Code of cities located within an area subjected to earthquake shocks."

After a severe earthquake in Long Beach, California, in 1933, everyone was shocked by the damage done in Long Beach, particularly the damage to school buildings. The only thing that prevented loss of life to schoolchildren was the fact that the earthquake occurred at 6:00 a.m. when schools were not in session. Following the Long Beach earthquake, the California state legislature passed the Riley Act, which required all buildings, with a few minor exceptions, be designed to resist a relatively small earthquake force. In addition, the Field Act was passed, which gave the State Division of Architecture jurisdiction over the structural design of all school buildings and the rules and regulations adopted to govern these designs. The counties and municipalities in the state also revised their building codes to include earthquake design requirements.

At the same time, the city of Los Angeles adopted a lateral force requirement of 8 percent of the sum of the self-weight plus half of the design live load. In 1943, Los Angeles enacted the first code requirement that related the lateral design force to the flexibility of the building. The lateral force coefficient was determined by the formula:

$$C = \frac{0.6}{N + 4.5} \tag{A.1}$$

where N = the number of stories above the one under consideration

Each story had to be designed to resist a lateral shear of C times the dead load above the story under consideration. However, the building height was limited to 13 stories, although this height limitation was later removed. Note that, for the ground story of a 13-story building,

$$C = 0.036$$

In 1948, a joint committee was formed by the Structural Engineers Association of Northern California (SEAONC) and the San Francisco section of the American Society of Civil Engineers (ASCE) to draft a model code for California building codes. Under the requirements of this code, the recommended base shear for design was

$$V = CW \tag{A.2}$$

where W = the dead load plus one-fourth of the design live load

The seismic coefficient was specified as

$$C = \frac{0.015}{T} \tag{A.3}$$

where T = the fundamental period of the building, which was determined by the formula:

$$T = \frac{0.05h}{D^{1/2}} \tag{A.4}$$

where h = the height of the building, in feet
D = the plan length of the building in the direction being considered, also in feet

In 1956, the city of San Francisco enacted a building code based on the joint committee recommendations but with the seismic coefficient increased to

$$C = \frac{0.02}{T} \quad 0.035 < C < 0.075$$

In 1957, the Structural Engineers Association of California (SEAOC) put forth the following recommendations:

$$V = KCW \qquad (A.5)$$

where K is a *framing factor* that was intended to introduce the effect of ductility into the building design through the design base shear. The value of K depends on the relative ductility capacity of the framing system, and four different types of construction were defined by the committee. These are as follows in order of increasing ductility capacity:

$K = 1.33$: "Box type" with vertical loads carried by bearing walls and shear walls
$K = 1.00$: Buildings having a vertical load space frame
$K = 0.80$: Buildings with a horizontal bracing system capable of resisting lateral loads
$K = 0.67$: Buildings with lateral force resisted by a moment-resisting space frame

The seismic coefficient was specified as

$$C = \frac{0.05}{T^{1/3}} \qquad (A.6)$$

The fundamental period:

$$T = \frac{0.05H}{D^{1/2}} \qquad (A.7)$$

where H = the height of the building, in feet
D = the length of the building parallel to the earthquake force considered

The 1961 edition of the UBC stipulated that the base shear be determined as

$$V = ZKCW \qquad (A.8)$$

where, due to the possible national application of this code, three seismic zones were established and Z was introduced as a zone factor having a value of 1 for Zone 3 and $1/4$ for Zone 1. The framing factors, K, were similar to those suggested by the 1957 SEAOC recommendations. The seismic coefficient was modified slightly to

$$C = \frac{0.05}{T^{1/3}} < 0.10 \qquad (A.9)$$

Note the upper bound of this recommendation is the same as the 1923 Japanese requirement. The fundamental period of vibration was estimated as follows:

$$T = 0.1N \text{ for moment-resisting frames}$$

$$T = 0.05H/D^{1/2} \text{ for other buildings}$$

The base shear was distributed over the height as lateral forces at the floor and roof levels as

$$F_x = V \frac{w_x h_x}{\sum_i w_i h_i} \qquad (A.10)$$

The overturning moment at the base was

$$M = J \sum F_x h_x \qquad (A.11)$$

where

$$J = \frac{0.5}{T^{2/3}} \qquad 0.33 \le J \le 1$$

The overturning moment, M_x, at level x above the base was to be determined by linear interpolation between the moment M at the base and zero at the top:

$$M_x = \frac{M(H - h_x)}{H} \qquad (A.12)$$

These recommendations take on new importance if one considers that the design life of a typical code-designed building is usually considered to be 50 years. This implies an aging population of buildings that reached their design limit in 2011. Evaluation of the seismic resistance of these older buildings will require knowledge of the seismic provisions at the time they were built and put into service. Some of these buildings will also have experienced one or more earthquakes during their lifetime, and the effect of these on the lateral resistance of the building will have to be evaluated.

In 1967, the UBC introduced the concept of ductile detailing of reinforced concrete moment frames and steel moment frames. It also included a requirement that a portion of the base shear force be applied as a concentrated load, F_t, at the top of the building. This modified loading was defined as follows:

$$F_t = 0 \quad \frac{H}{D} \leq 3$$
$$F_t = 0.004V \left(\frac{H}{D}\right)^2 \leq 0.15V \tag{A.13}$$

The remainder of the base shear was allocated to the various levels, as before:

$$F_x = (V - F_t)\frac{w_x h_x}{\sum wh} \tag{A.14}$$

This modification was intended to account for the effect of the higher modes of vibration in taller, more flexible structures. The base overturning moment was unchanged except that the extra roof force was now included:

$$M = J \left(F_t h_t + \sum_{1}^{n} F_i h_i\right) \tag{A.15}$$

In subsequent years, the overturning moment reduction factor was modified and then eliminated, thereby making $J = 1$.

In 1976, substantial changes were introduced into the earthquake regulations. A new seismic zone (Zone 4) was added, representing active fault regions in California and Nevada. The base shear formula was modified by adding two new factors as follows:

$$V = ZIKCSW \tag{A.16}$$

The new factor, I, is defined as an occupancy importance factor and is related to the potential hazard to life safety and varies between 1.0 and 1.5 based on occupancy:

$$1.0 \leq I \leq 1.5 \tag{A.17}$$

The upper bound of 1.5 is for essential facilities that must remain operational following an earthquake. An intermediate value of 1.25 is required for assembly buildings having an occupancy of more than 300 people in one room, and the lower bound of 1.0 is for all other buildings.

The factor, S, is a soil site resonance factor computed from the ratio of the building period to the soil site period as

$$
\begin{aligned}
S &= 1 + \frac{T}{T_S} - 0.5 \left(\frac{T}{T_S} \right)^2 \qquad \frac{T}{T_S} \leq 1 \\
S &= 1.2 + 0.6 \frac{T}{T_S} - 0.3 \left(\frac{T}{T_S} \right)^2 \qquad \frac{T}{T_S} > 1 \\
S &\geq 1.0
\end{aligned} \qquad \text{(A.18)}
$$

The code also specified that connections must either be designed to develop the full capacity of the members or else be designed for the code forces without the one-third increase usually allowed when earthquake forces are included in the design.

The seismic coefficient was modified slightly to

$$
C = \frac{1}{15T^{1/2}} \leq 0.12 \qquad \text{(A.19)}
$$

where $T = 0.10N$ for ductile moment frames
$\quad T = 0.05H/D^{1/2}$ for other buildings

These regulations also allowed the building period to be determined using the Rayleigh method in the form

$$
T = 2\pi \sqrt{ \frac{\displaystyle\sum_{i=1}^{n} w_i \delta_i^2}{g \displaystyle\sum_{i=1}^{n} F_i \delta_i} } \qquad \text{(A.20)}
$$

The lateral forces may be applied in any rational manner and not necessarily those obtained from code formulas. However, the displacements must be the displacements computed for the forces used. The product CS need not be taken to exceed 0.14.

As in the 1967 UBC, a portion of the base shear was required to be placed at the top of the building. However, this requirement was revised in the following manner:

$$
\begin{aligned}
F_t &= 0 & T &\le 0.7 \text{ sec} \\
F_t &= 0.07TV < 0.25V & T &> 0.7 \text{ sec}
\end{aligned}
\tag{A.21}
$$

Finally, the 1976 code included a drift limitation and a special provision for nonstructural elements. The calculated drift within any story caused by lateral forces is limited to the following:

$$
\begin{aligned}
\Delta &\le 0.005H & K &\ge 1 \\
\Delta &\le 0.005KH &
\end{aligned}
\tag{A.22}
$$

where Δ = the lateral displacement within a story

Most of the 1976 code provisions were carried over to the 1985 edition of the code.

A.2 1978 ATC 3-06 RESOURCE DOCUMENT FOR MODEL CODES

In 1978, a national group of experts in the field of earthquake engineering was convened by the Applied Technology Council (ATC) to write a resource document for model codes that would incorporate the latest thinking and research on the design of earthquake-resistant structures.

A.2.1 1985 NEHRP 85

Following review and comment, the main provisions of ATC 3-06 were released by the National Earthquake Hazard Reduction Program (NEHRP) as a code, NEHRP 85, with only minor changes. The design base shear was specified by the basic formula:

$$
V = C_s W
\tag{A.23}
$$

where W = the total weight of the structure, including permanent attachments

The seismic design coefficient is based on an earthquake having a 10 percent probability of exceedance in 10 years and is determined from the formula:

$$C_s = \frac{1.2A_v S}{RT^{2/3}} \leq \frac{2.5A_a}{R} \qquad \text{(A.24)}$$

where A_v = the effective peak velocity–related acceleration
A_a = the effective peak acceleration
S = the soil site coefficient
R = the response modification factor
T = the fundamental period of the building

The building period can be estimated using one the following formulas:

For moment-resisting frames:
$$T = C_T h_n^{3/4} \qquad \text{(A.25)}$$

where $C_T = 0.035$ for steel frames
$C_T = 0.030$ for reinforced concrete frames

For all other buildings:
$$T = \frac{0.05h_n}{\sqrt{L}} \qquad \text{(A.26)}$$

where L = the length of the building in the direction under consideration

The design interstory displacement, Δ, is the difference between the deflection δ_x at the top of the story under consideration and δ_x at the bottom of the story under consideration. The interstory displacement is based on the calculated deflections and is evaluated by the formula:

$$\delta_x = C_d \delta_{xe} \qquad \text{(A.27)}$$

where C_d = the given deflection coefficient ($1.25 \leq C_d \leq 6.5$)
δ_{xe} = the deflections determined by an elastic analysis

The deflection coefficient increases the elastic displacement caused by inelastic behavior, whereas the response modification factor reduces the force caused by inelastic behavior.

A.2.2 1988 UBC

This version of the UBC provides for the use of the equivalent static force procedure or a dynamic analysis for regular structures under 240 ft in height but requires a dynamic analysis for irregular structures and structures over 240 ft tall.

The base shear was given by the formula:

$$V = \frac{ZICW}{R_W} \tag{A.28}$$

where Z = the seismic zone factor

I = the occupancy importance factor

C = the numerical coefficient, which is dependent on the soil conditions at the site and the period of the structure

W = the dead load of the structure

R_W = a factor that represents the ductility of the structural system

The seismic site coefficient, which depends on the soil characteristics and the fundamental period of the structure, is specified by the formula:

$$C = \frac{1.25S}{T^{2/3}} \leq 2.75 \tag{A.29}$$

$$\frac{C}{R_W} \geq 0.075 \tag{A.30}$$

The building period may be determined by either analysis or by using an empirical formula. The empirical formula has the form

$$T = C_t h_n^{3/4} \tag{A.31}$$

where C_t = 0.035 for steel moment frames

C_t = 0.030 for reinforced concrete moment frames and eccentric braced frames

C_t = 0.020 for all other buildings

If the period is determined using Rayleigh's method or another method of analysis, the value of C must be at least 80 percent of the value obtained by using the appropriate empirical formula.

The interstory drift for buildings that are 65 ft or taller cannot exceed either of the following two constraints:

$$\Delta \leq \frac{0.03h}{R_W}$$

(A.32)

$$\Delta \leq 0.004h$$

For buildings less than 65 ft in height, the interstory displacement cannot exceed either of the following:

$$\Delta \leq \frac{0.04h}{R_W}$$

(A.33)

$$\Delta \leq 0.005h$$

A.2.3   1997 UBC

The 1997 revision to the UBC incorporates many of the recommendations of ATC 3-06. Although the code provides for the use of the equivalent static force procedure, it expands the requirement of a dynamic analysis to include irregular structures over 65 ft in height and buildings located on poor soils (type S_F) and having a period greater 0.7 sec. In this edition, emphasis is placed on strength design rather than working stress, and the seismic loads are based on the strength condition. This is a departure from previous editions of the UBC, which were based on working stress design.

Design Base Shear   The design base shear is specified by the formula:

$$V = \frac{C_v I}{RT} W$$

(A.34)

where T = the fundamental period of the structure in the direction under consideration

I = the occupancy importance factor

C_v = the velocity-related seismic coefficient, which now includes the effect of the general soil conditions at the site

W = the seismic dead load

R = a factor that accounts for the ductility and overstrength of the structural system

The base shear specified by this equation is subject to the following three limitations:

1. The design base shear need not exceed

$$V = \frac{2.5C_a I}{R} W \tag{A.35}$$

2. The design base shear cannot be less than

$$V \geq 0.11C_a IW \tag{A.36}$$

where C_a = the acceleration-related seismic coefficient, which also includes the effect of the general soil conditions at the site
W = the seismic dead load, as before
I = the occupancy importance factor

3. In the zone of highest seismicity (Zone 4), the base shear requirement is

$$V \geq \frac{0.8ZN_v I}{R} W \tag{A.37}$$

where Z = the seismic zone factor
N_v = the near-fault source factor and the other factors are as defined previously.

Building Period   The basic formula for the building period is the same as that in the 1988 UBC. The exception is that the constraints on the use of Rayleigh's method have been relaxed. In Zone 4, the value calculated using a Rayleigh procedure cannot be more than 30 percent that given by the formula. In the other three seismic zones, it cannot be greater than 40 percent.

Drift Limitations   The maximum inelastic displacement is defined as

$$\Delta_M = 0.7R\Delta_S \tag{A.38}$$

This displacement must include both translation and torsion along with elastic and inelastic contributions. The displacement, Δ_S, is the design-level response displacement. For structures with a period less than 0.7 sec, the maximum story drift is limited to

$$\Delta_a \leq 0.025h \tag{A.39}$$

where $h =$ the story height

For structures with a period greater than 0.7 sec,

$$\Delta_a \leq 0.020h \tag{A.40}$$

A.3   2000–2009 INTERNATIONAL BUILDING CODE (IBC 2000, 2003, 2006, AND 2009)

Starting with the year 2000, the local code development entities in the United States joined forces to develop and publish a national model code titled the "International Building Code," or IBC. This building code takes its provisions from the latest NEHRP provisions for design of new buildings. The IBC incorporates the seismic design provisions of the ASCE 7 standard by reference with some modifications called *amendments*. The regional and local codes are developed by adopting the IBC and amending it, as needed, to reflect their specific needs. For example, the California Building Code (CBC) is an amended version of the IBC, and the Los Angeles Building Code is an amended version of the CBC. Although specific provisions contained in the IBC have developed and changed from IBC 2000 to IBC 2009, the general format and approach have basically stayed the same. The seismic analysis requirements of ASCE 7-05, which are incorporated by reference into IBC 2009, are presented in Chapter 9 of this book.

SELECTED REFERENCES

Berg, Glen V. 1989. *Elements of Structural Dynamics*. Englewood Cliffs, NJ: Prentice Hall.

Biggs, John M. 1964. *Introduction to Structural Dynamics*. New York: McGraw-Hill.

Chopra, Anil K. 2007. *Dynamics of Structures*, 3rd ed. Upper Saddle River, NJ: Prentice Hall.

Clough, Ray W., and Joseph Penzien. 1975. *Dynamics of Structures*. New York: McGraw-Hill.

Green, Norman B. 1978. *Earthquake Resistant Building Design and Construction*. New York: Van Nostrand Reinhold.

Magrab, E. B., S. Azarm, B. Balachandran, J. H.Duncan, and G. C.Walsh. 2011. *An Engineer's Guide to MATLAB*, 3rd ed. Upper Saddle River, NJ: Prentice Hall.

MathWorks, Inc. 2011. *Learning MATLAB 7*. Natick, MA: MathWorks, Inc.

McCuskey, S. W. 1959. *An Introduction to Advanced Dynamics*. Reading, MA: Addison-Wesley.

Naeim, Farzad. 1989. *The Seismic Design Handbook*. New York: Van Nostrand Reinhold.

_____. 2001. *The Seismic Design Handbook*, 2nd ed. Boston: Kluwer.

INDEX

US Geological Survey (USGS), 212
USGS Hazard Calculator,
213–215

Vehicle loads, 2
Velocity, relative, 15, 22, 205
Virtual displacement:
defined, 8
single-degree-of-freedom systems,
20–23
Virtual work:
principle of, 8

single-degree-of-freedom systems,
20–23
Viscous damping:
harmonic loading, 84
multiple-degree-of-freedom systems,
139–140

Weight, mass vs., 6
Wind loads, 2–3
Work:
concept of, 6–7
virtual, 8, 20–22